WEDNESDAY AT THE PIER

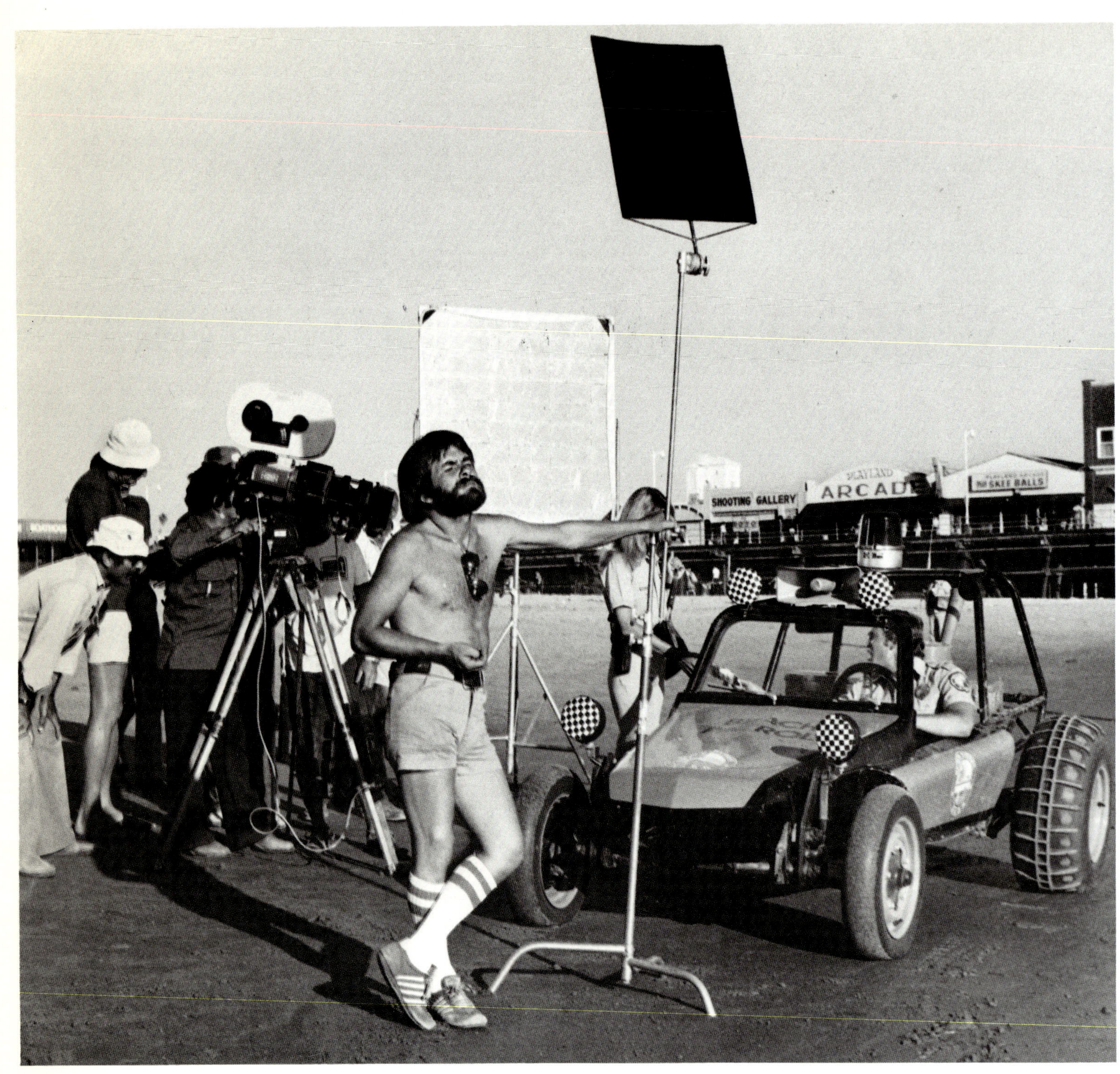

1. Filming "Beach Patrol"

WEDNESDAY AT THE PIER

Visitors and Regulars
at the Santa Monica Pier

Photographs and Text
by
Tom Zimmerman

Shapolsky Publishers
New York

A Shapolsky Book

For any additional information, contact:
Shapolsky Publishers, Inc.
136 West 22nd Street, New York, New York 10011

First Printing

ISBN 1-56171-084-2

Cover Design by Jim Heimann

For Anthony Turbollow
who taught me all about history and life

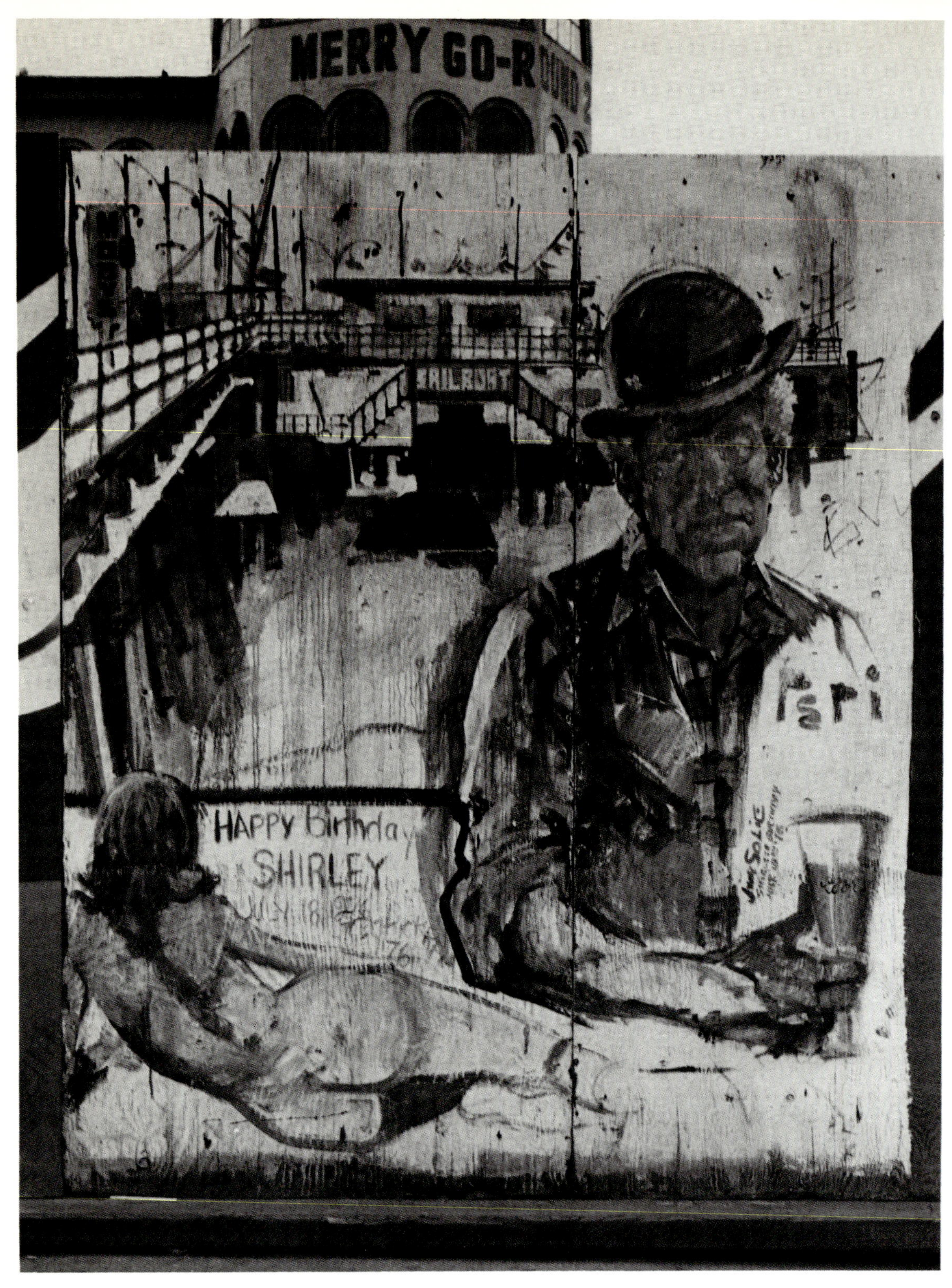

2. Wall Painting, Boardwalk

ACKNOWLEDGMENTS

Wednesday at the Pier has had the benefit of the thoughtful criticism of lots of people. T.G. Lane offered advice right from the start. My brother, Paul Zimmerman, was always full of ideas as he worked on a separate project in Venice, California. Monique LaVoie and Kathryn Burbach were very good about not only posing for dozens of pictures, but in sharing their insight on the Pier.

Down at the Pier, Wild Bill, Crispy Cory, Al the Photographer, and Slim Jim all helped me get to know the regulars they spent so much of their time with. Laura Lopez introduced me to many of the other carny people. She's a wonderful person. It inevitably made my day to talk to her.

As always, Michael John Sullivan was encouraging and patiently critical, offering insightful comments that helped greatly to improve the manuscript. Nancy Newman Work never failed to cheer me on and suggest ways to improve photos.

The winnowing process of images was aided by all of *Wednesday at the Pier's* showings. I want to thank Susan Weinberg of the Infinity Gallery, Mitzi Landau of the Landau Gallery, Lee Draper of the Santa Monica Heritage Square Museum, and Molly Jensen of the Tampa Museum, Barry Brennan of the *Santa Monica Outlook* and Bill Rollins of the *Los Angeles Times.* Thomas Hines and Rachel Lozzi also took the time to critically evaluate the series at these earlier exhibits.

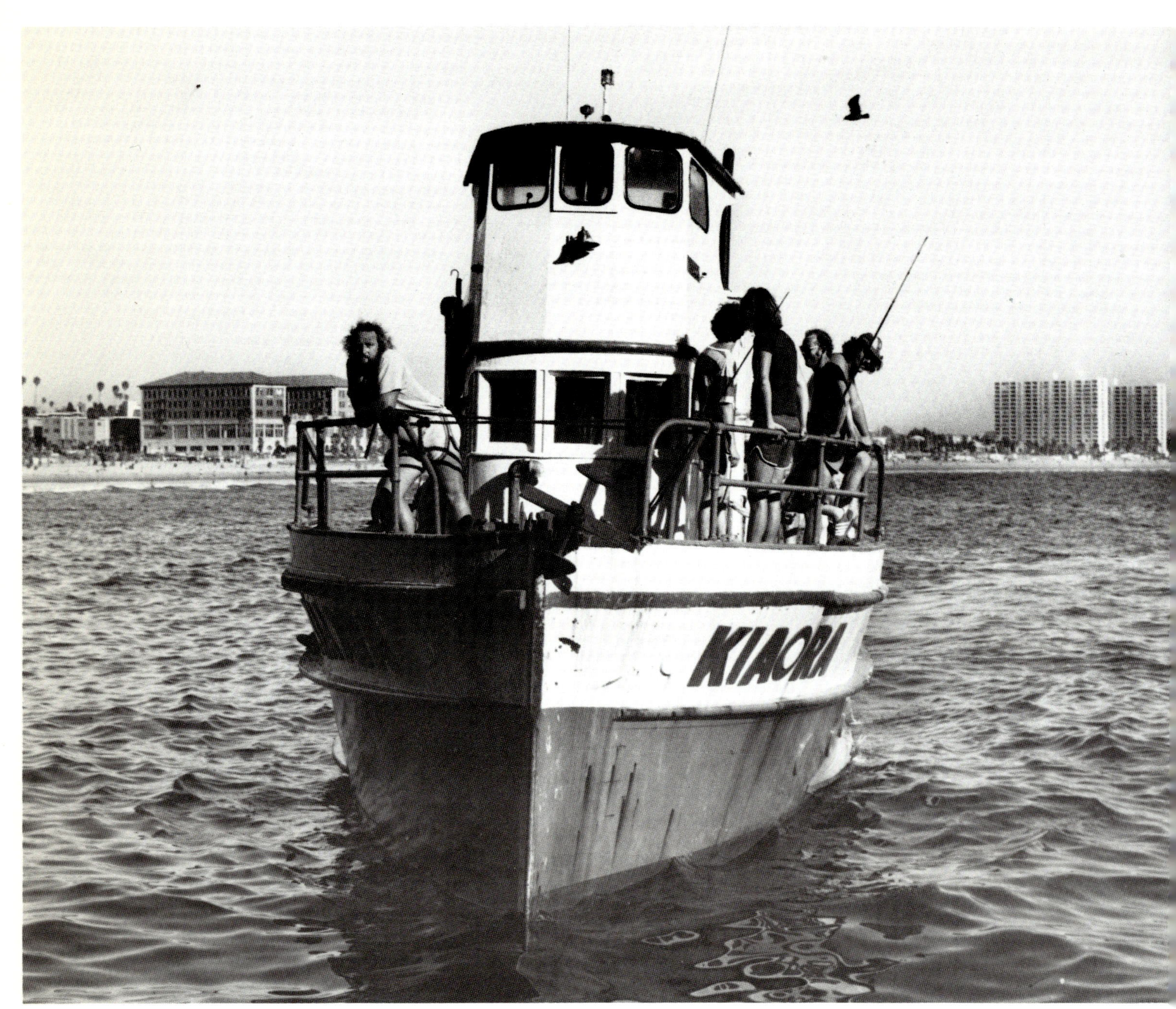

3. Kiora *Docking*

CONTENTS

I: INTRODUCTION

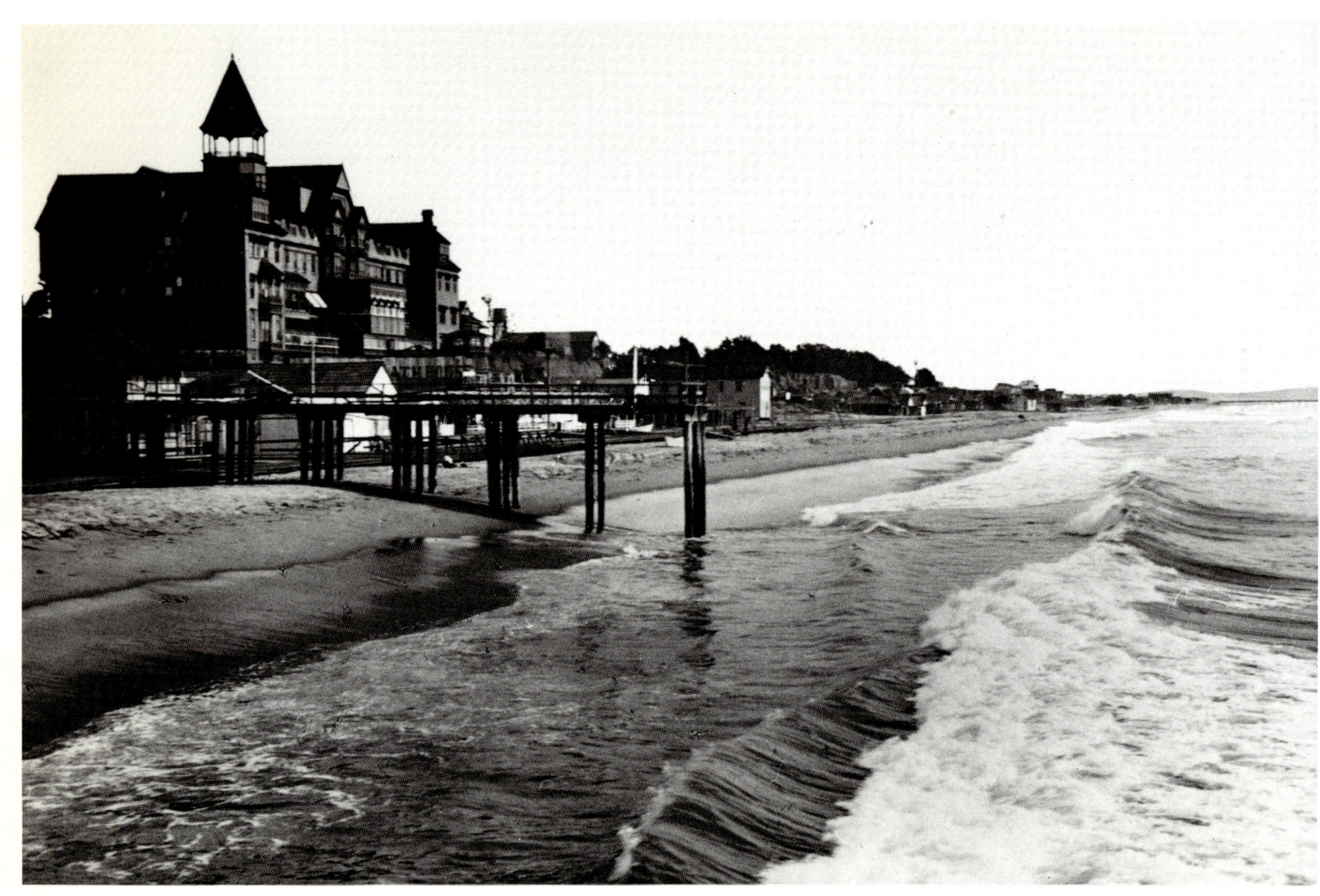

4. Arcadia Hotel, 1890

The Pleasure Piers. Seaside fun zones. Roller coasters and merry-go-rounds. Carny games of all sorts and dance halls. Movie theaters and food. And, of course, the ocean surrounding the whole affair. San Diego had Belmont Shores, Long Beach had the Pike. Los Angeles had the huge fun zones at Venice and Ocean Park, and a much smaller one at Santa Monica.

It was late fall. Cool air. Alternating dark and white clouds. I had a new job that would start next week, so I could relax today. Drove north through the Malibu hills taking pictures. Coming back to L.A. on the Pacific Coast Highway, I stopped at the Santa Monica Pier. Hadn't been there since I was a little kid and my mother took me to the merry-go-round—as her mother had done with her. A walk out into the sea. All of the city's ethnic groups and cultures passing by. People fishing, others just hanging out or playing pin ball in the Arcade. World War II newspaper headlines on a partition in the carousel, while recorded caliope music played. Time's continuum.

I started making a serious effort to photograph what I saw at the Pier in 1978. The photos were all taken on Wednesday because a day in the middle of the week offered the best chance to know the people who called the Pier home while still seeing plenty of tourists during the summer months. I wondered how the seasons affected the Pier. It was packed all summer long, but what of the rest of the year?

I went to the Pier almost every Wednesday for three years, and returned off and on afterwards whenever there was enough time. I met a lot of the tourists that came to the Pier as part of their visit to Los Angeles, but most interesting to me were the people who made the Pier their community.

Of all the things desperately lacking in late 20th Century American life, the sense of community tops the list. A need to belong to something larger than yourself is as certain a human need as food. Otherwise, you wind up one of the vacant men, aimlessly prowling the underbelly of the country, looking for ways to exorcise your

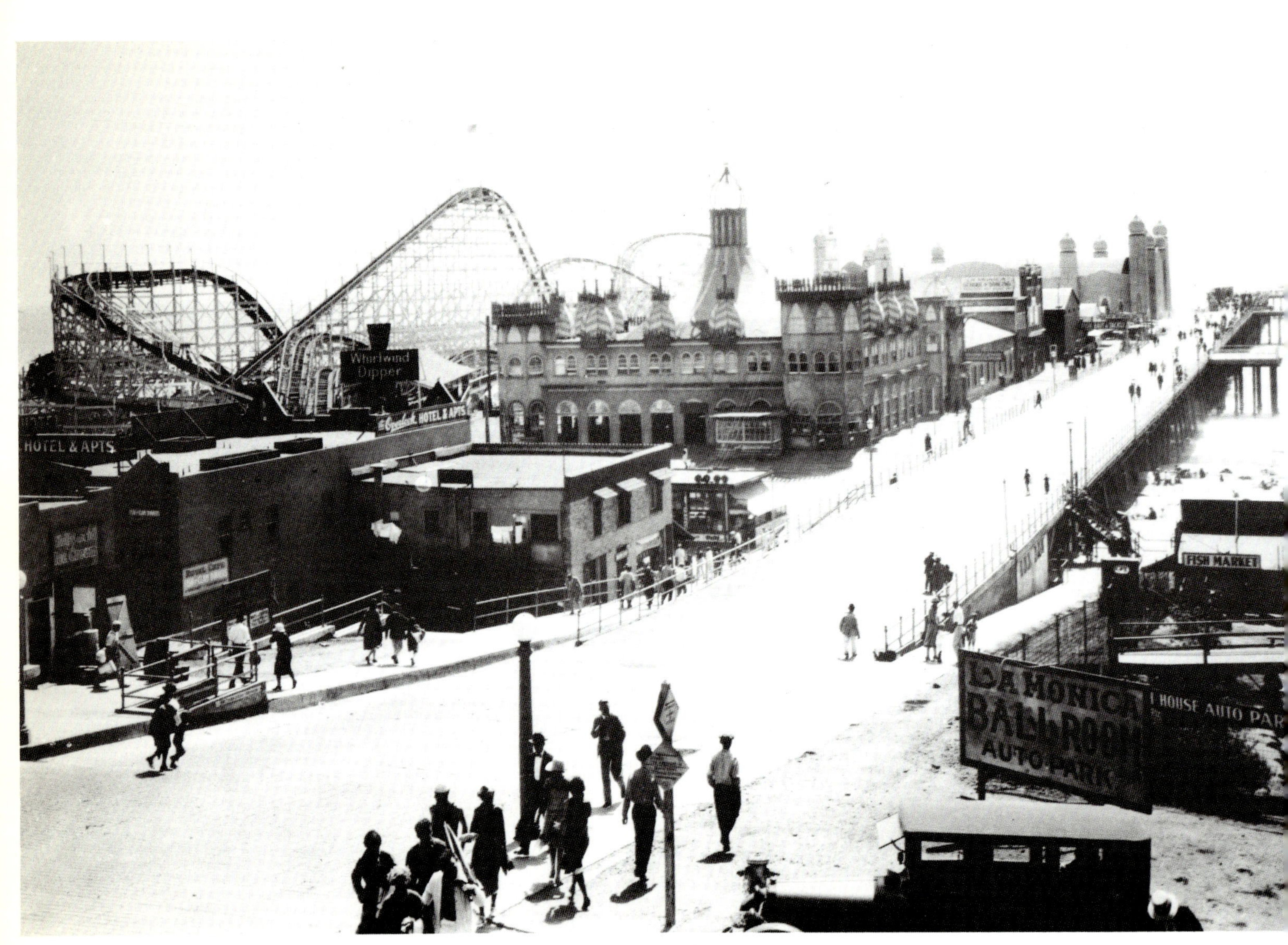

5. Santa Monica Pier, 1926

excruciating disconnectedness. The people that called the Pier home were not the dedicatedly upwardly mobile folks so at home on the refurbished Main St. in nearby Ocean Park. But neither were they hollow people, staring out with dead eyes at the world going by them. They had developed their own community.

This, of course, was not what the city government of Santa Monica had in mind for its waterfront. There was always a social factor about the Santa Monica beachfront lacking elsewhere on the L.A. area coast. Ocean Park and Venice were pure carny once Abbott Kinney was no longer around to try and instill culture. Santa Monica, on the other hand, had eleven beach clubs and the multi-million dollar Marion Davies estate near its Pier.

By the early 1960s all this was coming unglued. Marion Davies was long since moved out, most of the beach clubs were closed, and the Pier had disolved into seediness. The broader horizons post-war affluence offered the middle class and the popularity of Disneyland and the other theme parks spelled the end of the Pleasure Piers and all the other in-city diversions that had been so popular before World War II. The Ocean Park Pier managed to delay the end by becoming Pacific Ocean Park. But it only prolonged the agony.

The theme parks were so safe. Guarded gates, an army of workers to sweep up any trash that found its way onto the ground. Organized, sanitized. A carefully manipulated experience of fun. None of the chaos of the piers, those permanently placed cousins of the traveling carnival shows. They relied on the magic of seashore, painted wood, the inspired spiel of barkers, and most importantly, the imagination of the audience to create an atmosphere. The artifice is so transparent by the light of the sea air.

The 1960s disolved into the 1970s. The Marina del Rey had been opened by then. New housing and apartment developments sprouted up right along with it. The Main St. slum in Ocean Park was being redeveloped into shops, boutiques, and restaurants for affluent

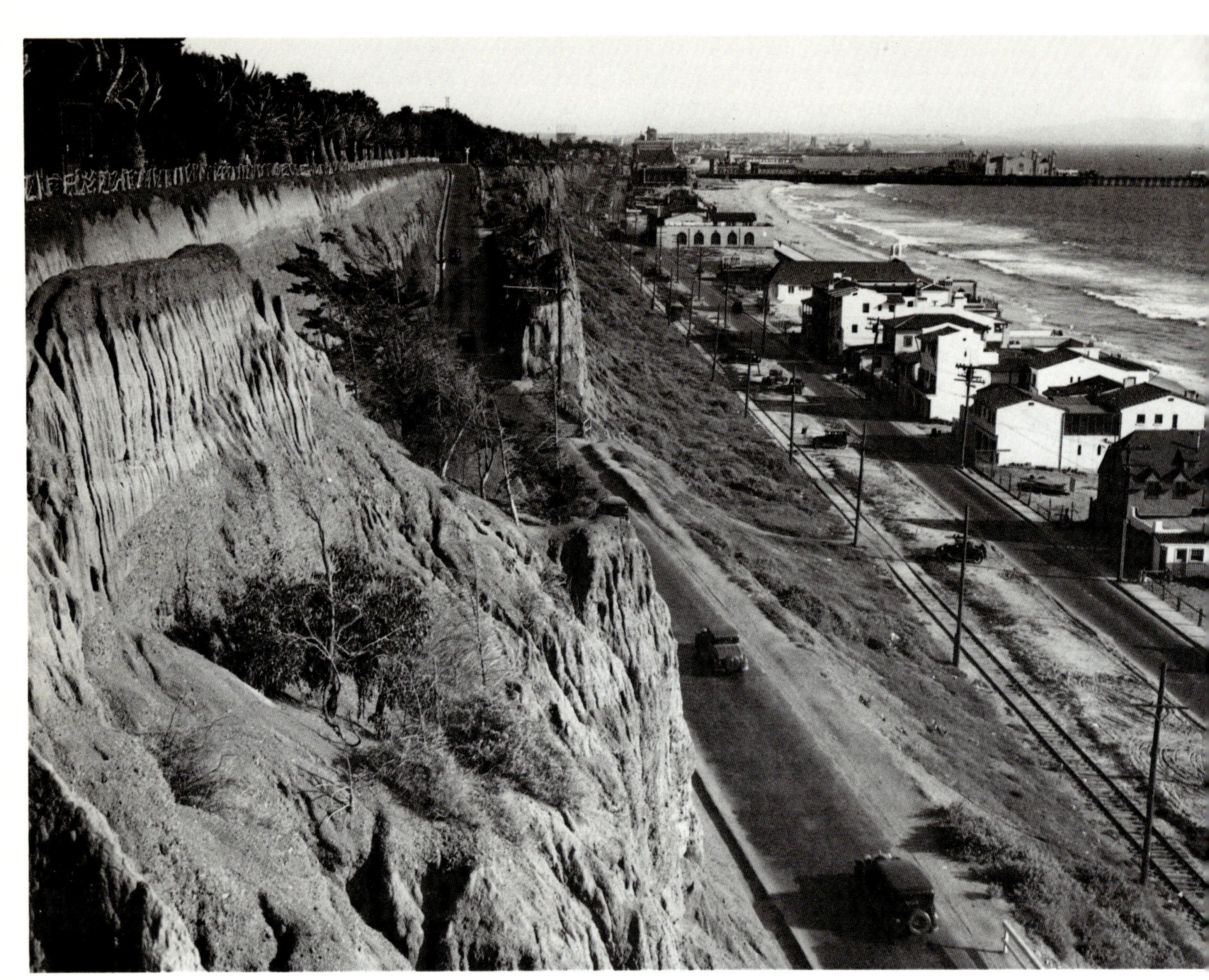

6. The Santa Monica Palisades, 1931

Westsiders. Even Venice, home to beatniks and hippies in their respective flowerings, and artists at all eras, was being gentrified. Serious money was flowing into the coastal areas.

This boded ill for the continued existence of the Pier. Bus lines and freeways from all over Los Angeles converge on it. The Pier remains one of the best cheap dates in the city. For a few dollars you can play pinball or carny games. Hang out on the sand. Ride the bumper cars. Meet girls or guys. The bumper cars are great for that. The etiquette of bumpering as courtship involves a fellow slamming into the car of the girl who strikes his fancy. What transpires after that is up to the girl.

All the major attractions had disappeared by the late 1970s. The Whirlwind Roller Coaster was closed in 1930. The weight lifting platform at Muscle Beach was removed in 1958. The already weakened breakwater that created a small boat harbor was almost destroyed by storms in 1959 and never attracted many yachts thereafter. The La Monica Ballroom was demolished in 1962 after ending its life as a skating rink.

All that was left was the bracing ambience of the sea rolling past wooden pilings, fishermen patiently waiting for their meals to swim by, a merry-go-round, bumper cars, several carny games, fast food joints, and two nice restaurants in the Boathouse and Moby's Dock. All this was yours for just showing up. Never an admission charge. No guards in period costumes roaming around. Just a collection of buildings from L.A.'s less structured past.

The original impetus for the city of Santa Monica to build a municipal pier at the foot of Colorado Blvd. was sewage. It had been using the outfall pipe beneath the Ocean Park Pier whose capacity was rapidly becoming overloaded. In 1909 the city opened its own 1600 foot long to stroll on, fish off, and hold a very long pipe caring effluence into the waters of the Pacific.

In 1916, experienced carousel operator and creator, Charles Looff convinced the city to permit him to build a Pleasure Pier next to the

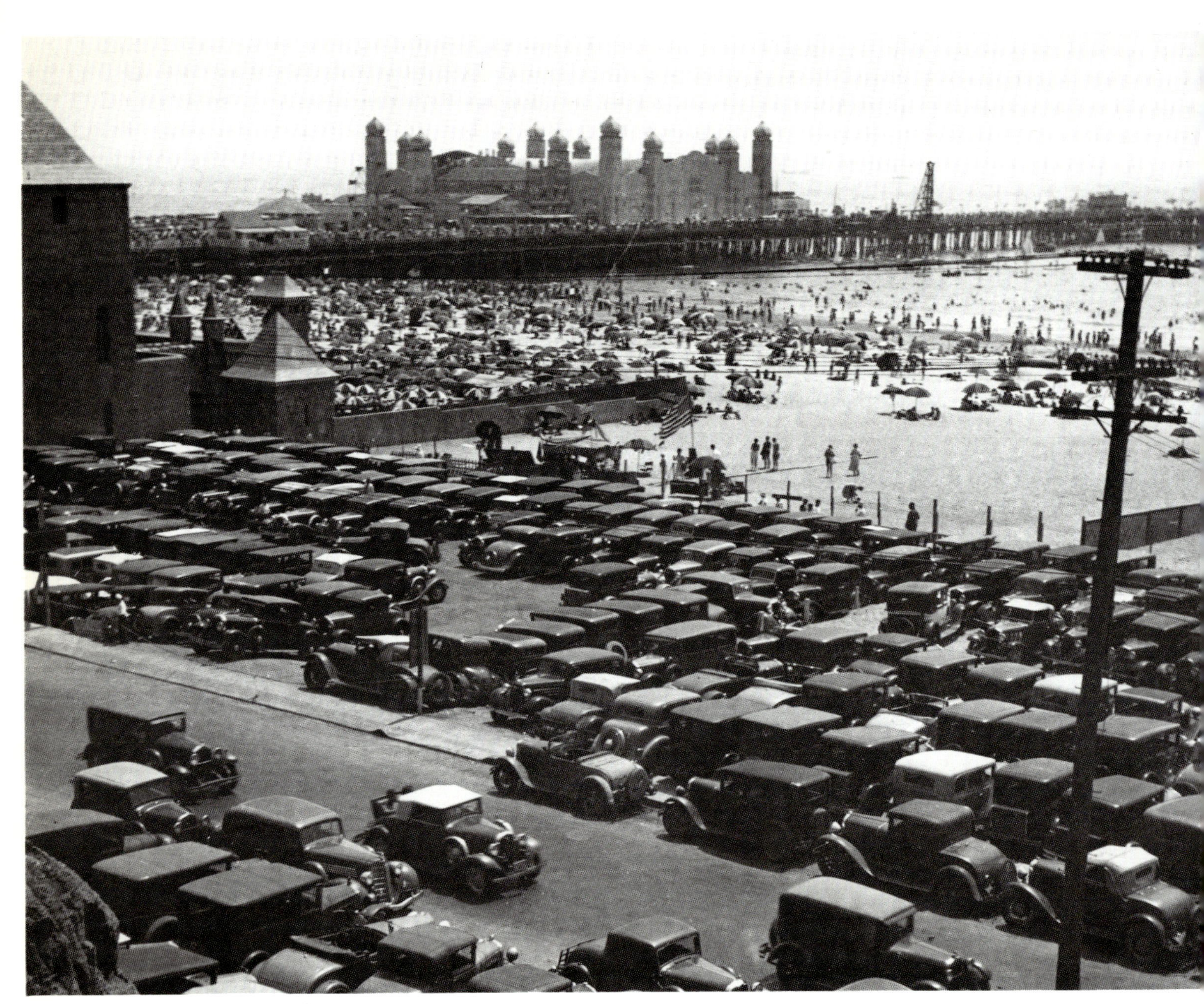

7. *The Glorious 4th, 1933*

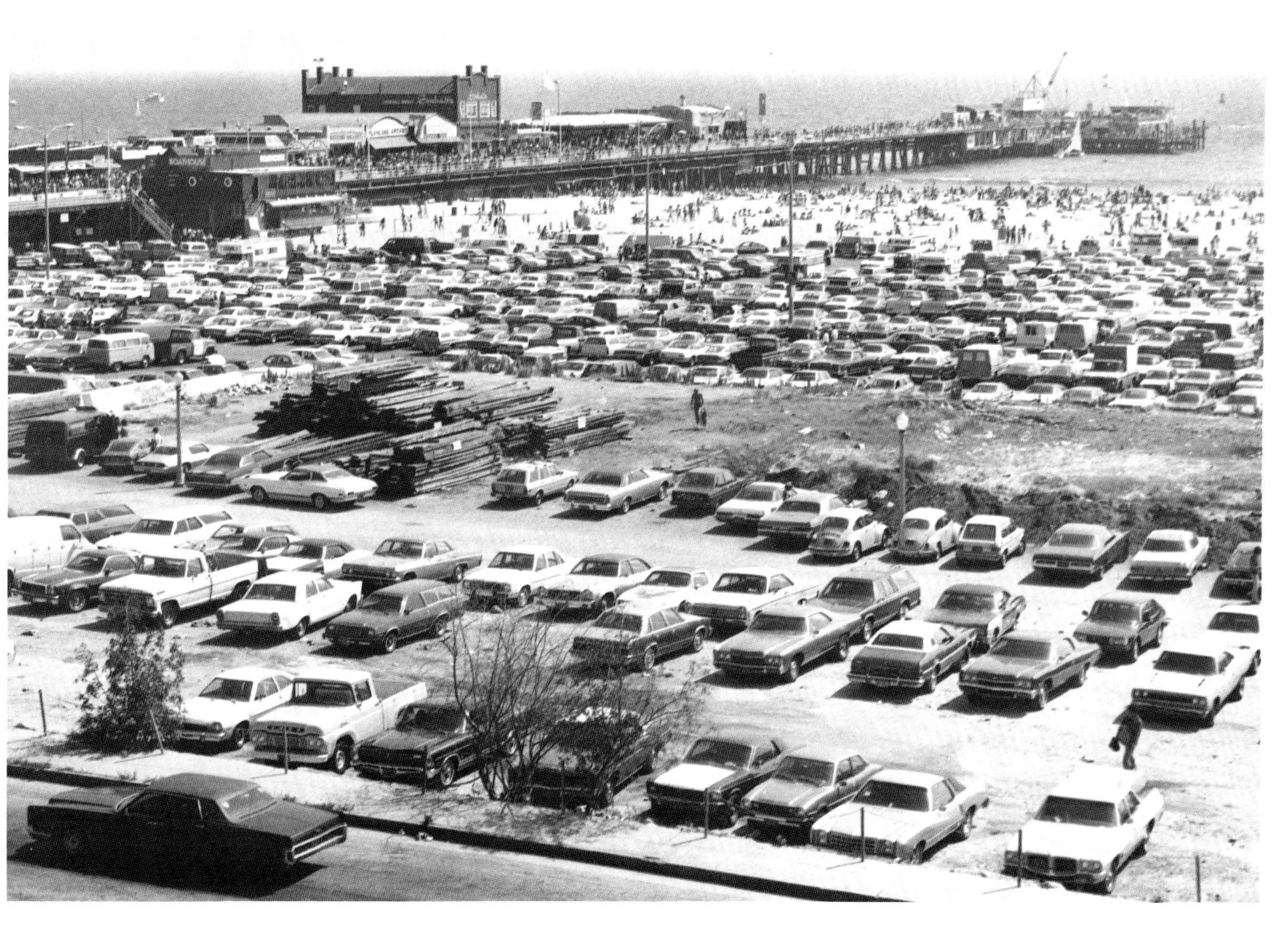

8. The 4th of July, 1979

Municipal Pier. Permission was granted and Santa Monica soon had a Pier with the games expected of a seaside resort.

The Looff Pier (later called the Newcomb Pier after its purchase by new owners in 1943) was in the American amusement park tradition. Our version of such affairs was radically different from their European antecedents. In Western Europe, the first amusement parks had been set up for the upper classes and were places to go for some rest. Lovely parks with manicured surroundings. This would never do for the United States with its democratic codes and combative social landscape. The parks on our side of the Atlantic were not built for relaxation. The rides in them were meant to scare, challenge, and unsettle. As a nation, our strength has always been aggression, never contemplation.

By the 1970s, Santa Monica had found other ways to dispose of its sewage. The city government was also in the market for disposing of its Pier. The most bizarre idea was published in 1972. It called for building an island in Santa Monica Bay bristling with luxury hotels, restaurants, and shops. Sufficient funds could thankfully never be raised and the plan died.

A few years later, it was supplanted with a scheme to radically upgrade the attractions on the Pier. By this time all the anti-Pier city council members had been voted out of office. People in Santa Monica obviously wanted to keep their Pier, but they just weren't too sure what they wanted done with it.

The storms of the Winter of 1983 almost decided the issue for the city. The Pier was halved, and what was left demanded immediate attention. The previous Summer, Susan Mullin, the city's representative on the Pier, helped organize a series of meetings to determine what the nature of the Pier should be. The message was always the same. "Basically the citizens have said we want this Pier fixed up. But keep the atmosphere the same." Mullin had her work cut out for her after the storms.

Today, the area around the Pier's entrance on Ocean Ave. has certainly been upgraded. Shops, restaurants, and boutiques have sprouted all over the place. But the Pier is its same old unique self. A good deal of paint has been slapped on the buildings and the old signs from the horses thanking people for saving them are gone from the Carousel Building, but the atmosphere has hardly changed at all.

The major attraction remains the gentle meeting of sea and land. There is some sort of deep seated race memory in humans that is lightly stroked by walking out over the Ocean. The safe crossing of water is calming to us. Throw in a few carny games, the best arcade in L.A., and the most sumptuous merry-go-round in Southern California and you have a fine way to spend part of your day.

May McCabe, who has worked on L.A. Pleasure Piers since before World War II, summed it up best. After forty years of making the same journey down to the waterfront, she still looked forward to getting down to the Pier. "I'll tell you one thing," she said, "no matter how rotten you may feel when you get up in the morning, when you come down to the Pier the feeling just disappears. This place is a tonic. It surely is."

II: THE PIER

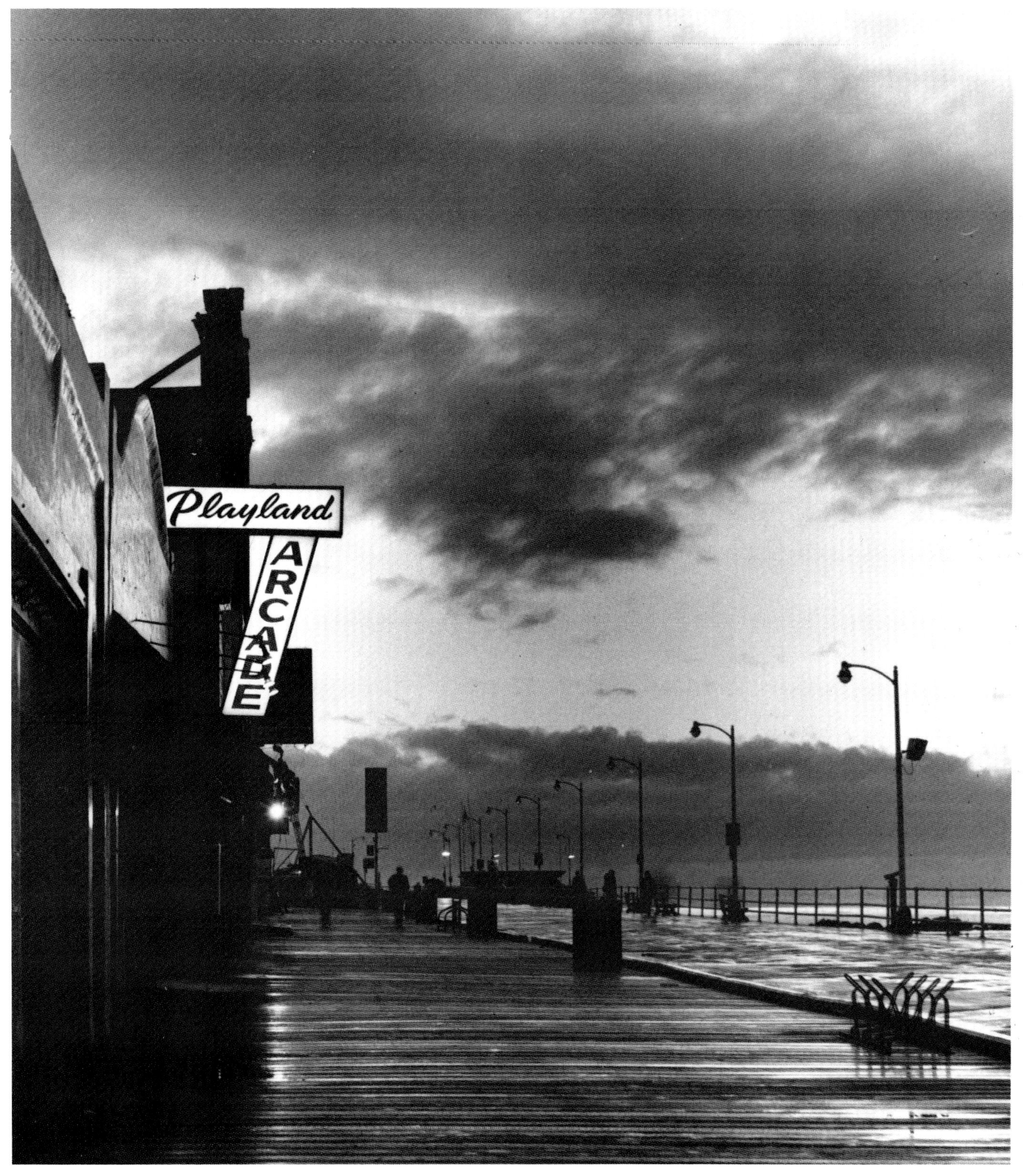

9. Clearing Storm

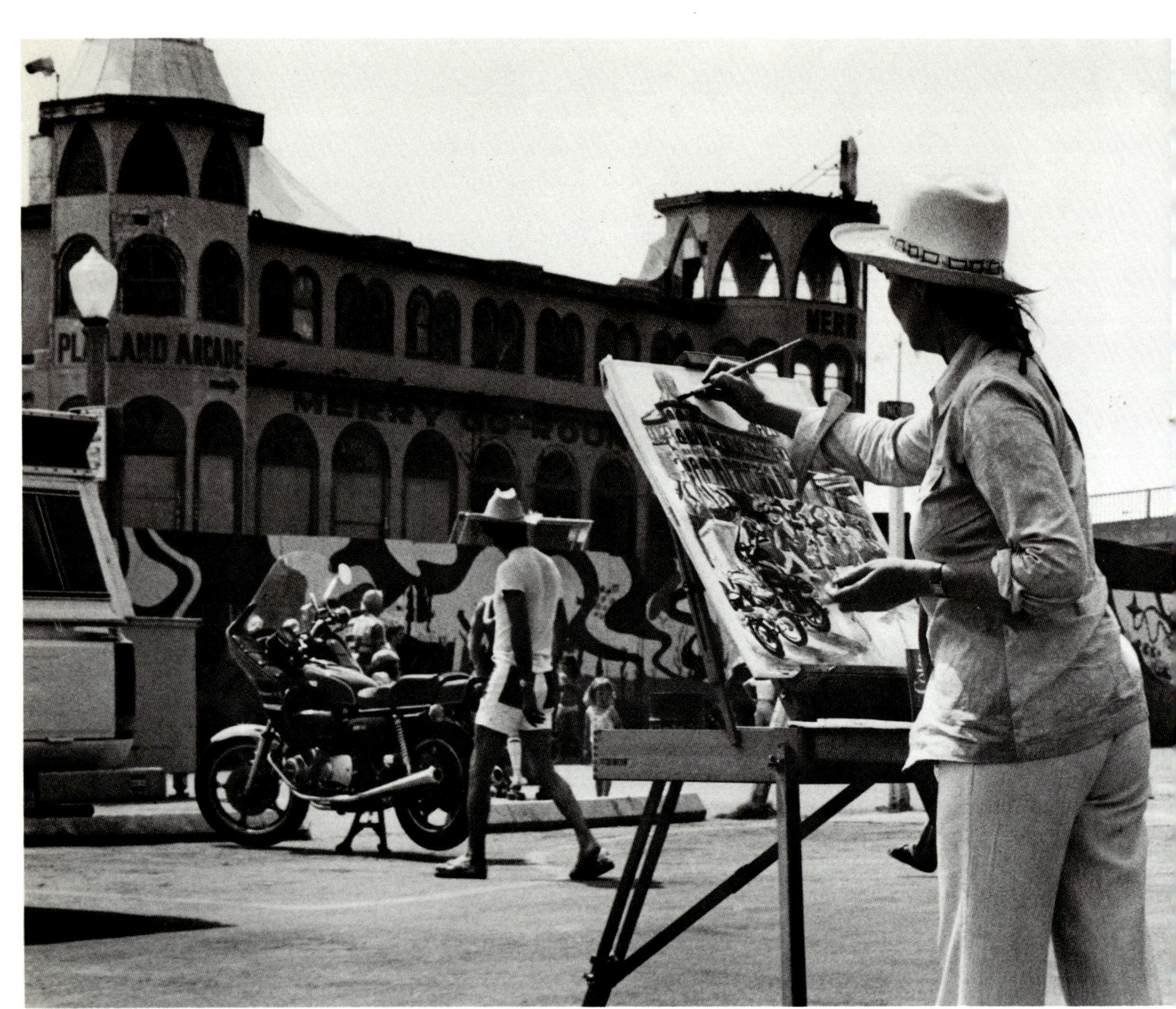

10. Boardwalk Painter

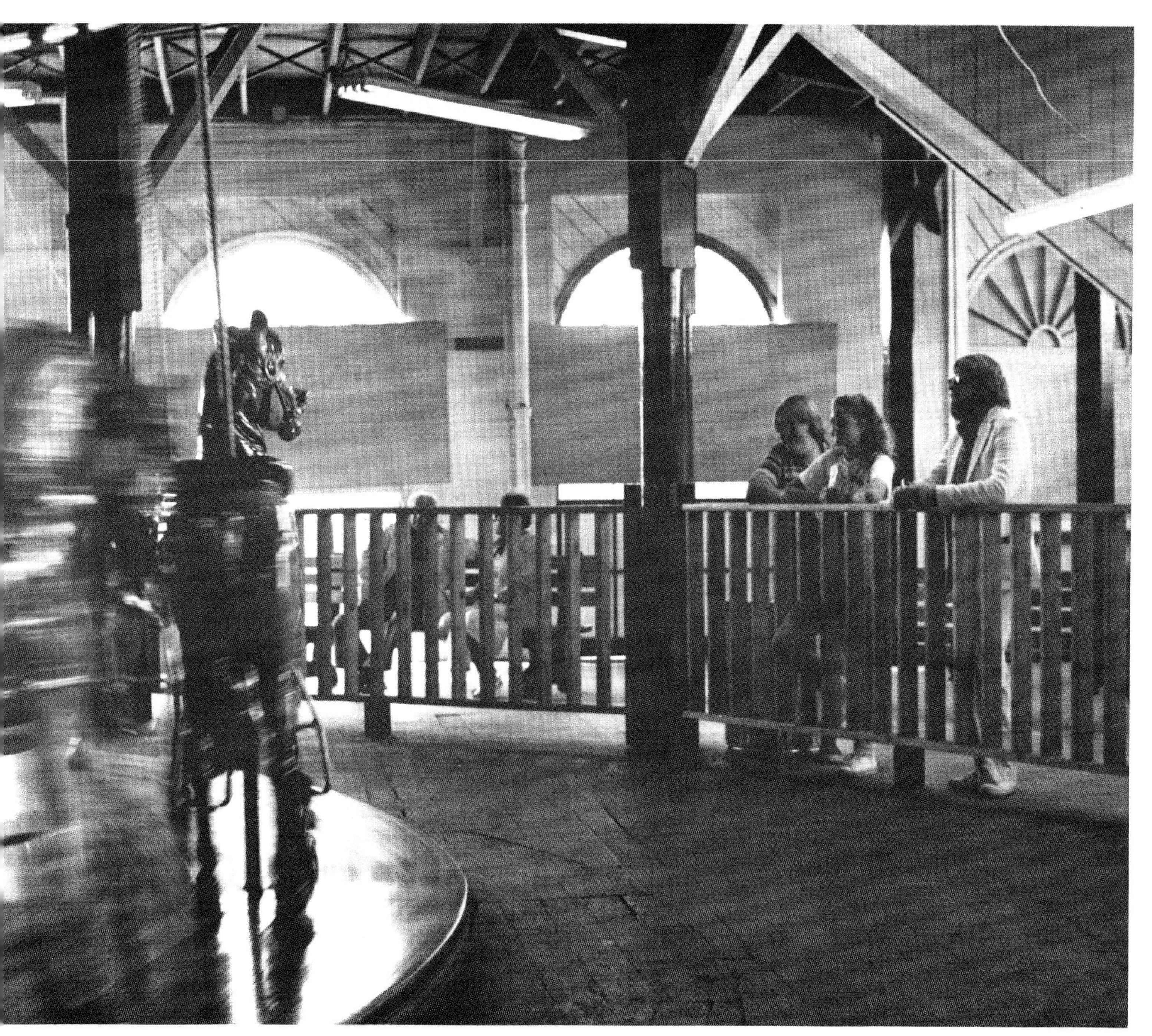

11. Merry-Go-Round

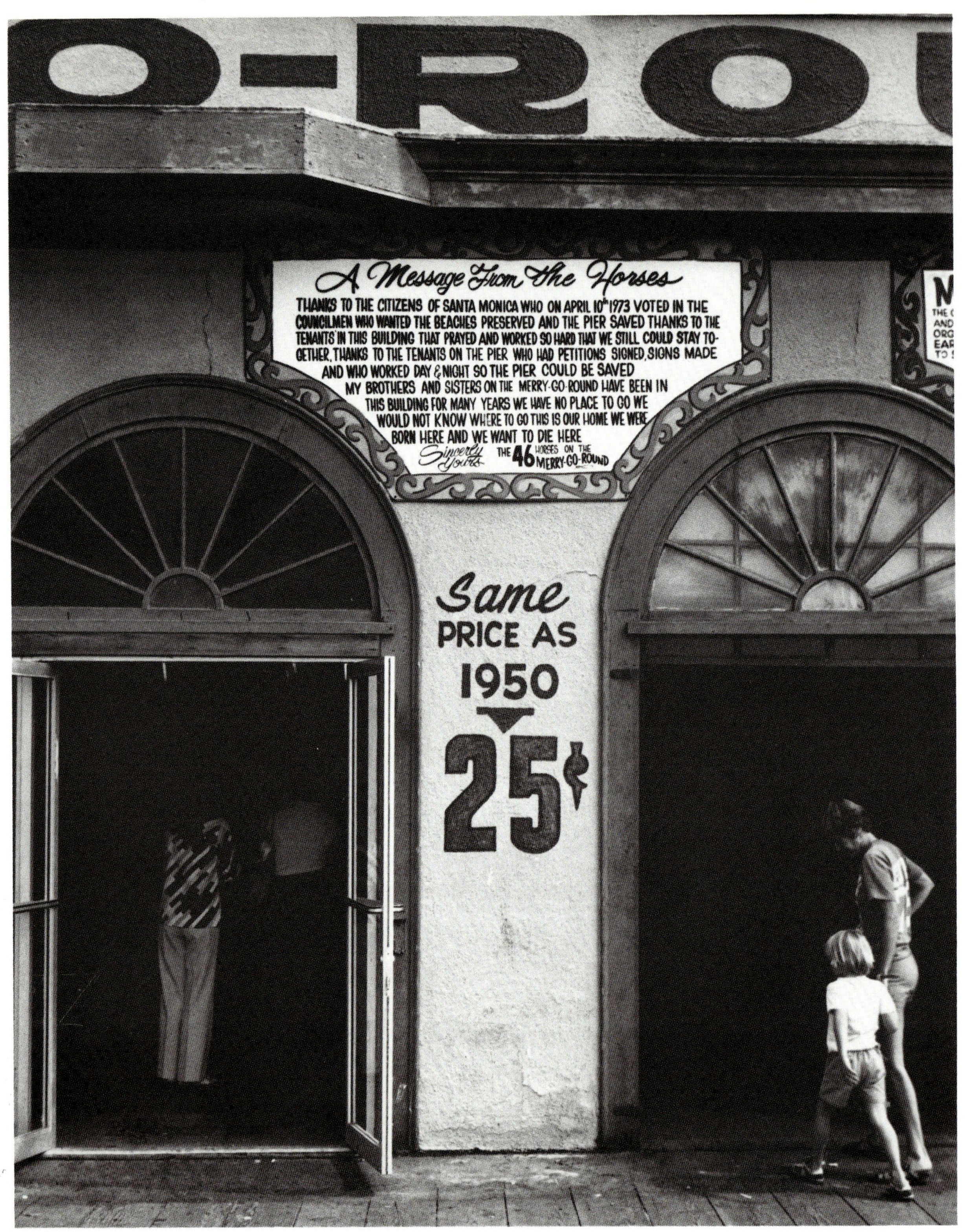

12. Carousel Sign

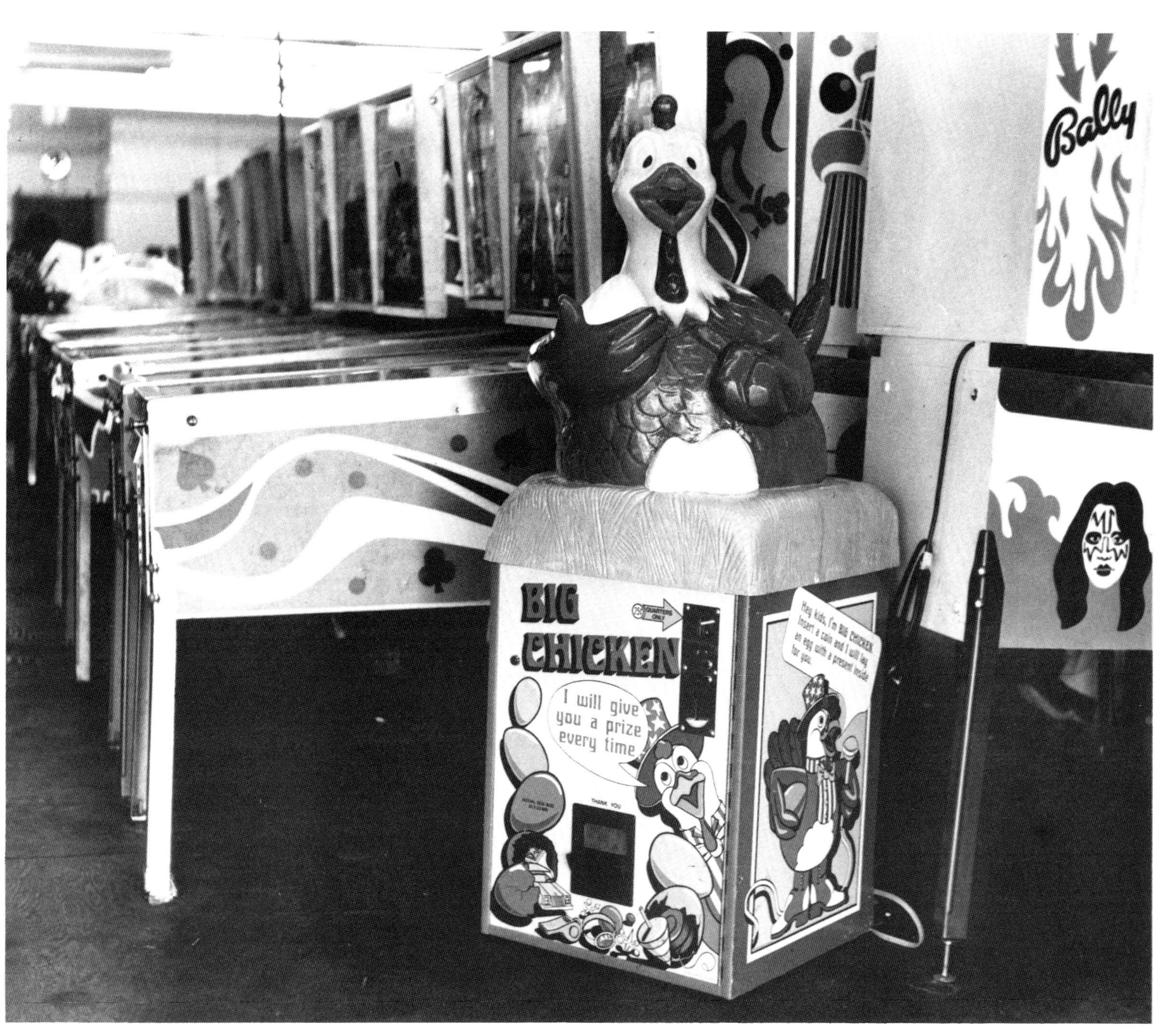

13. Big Chicken

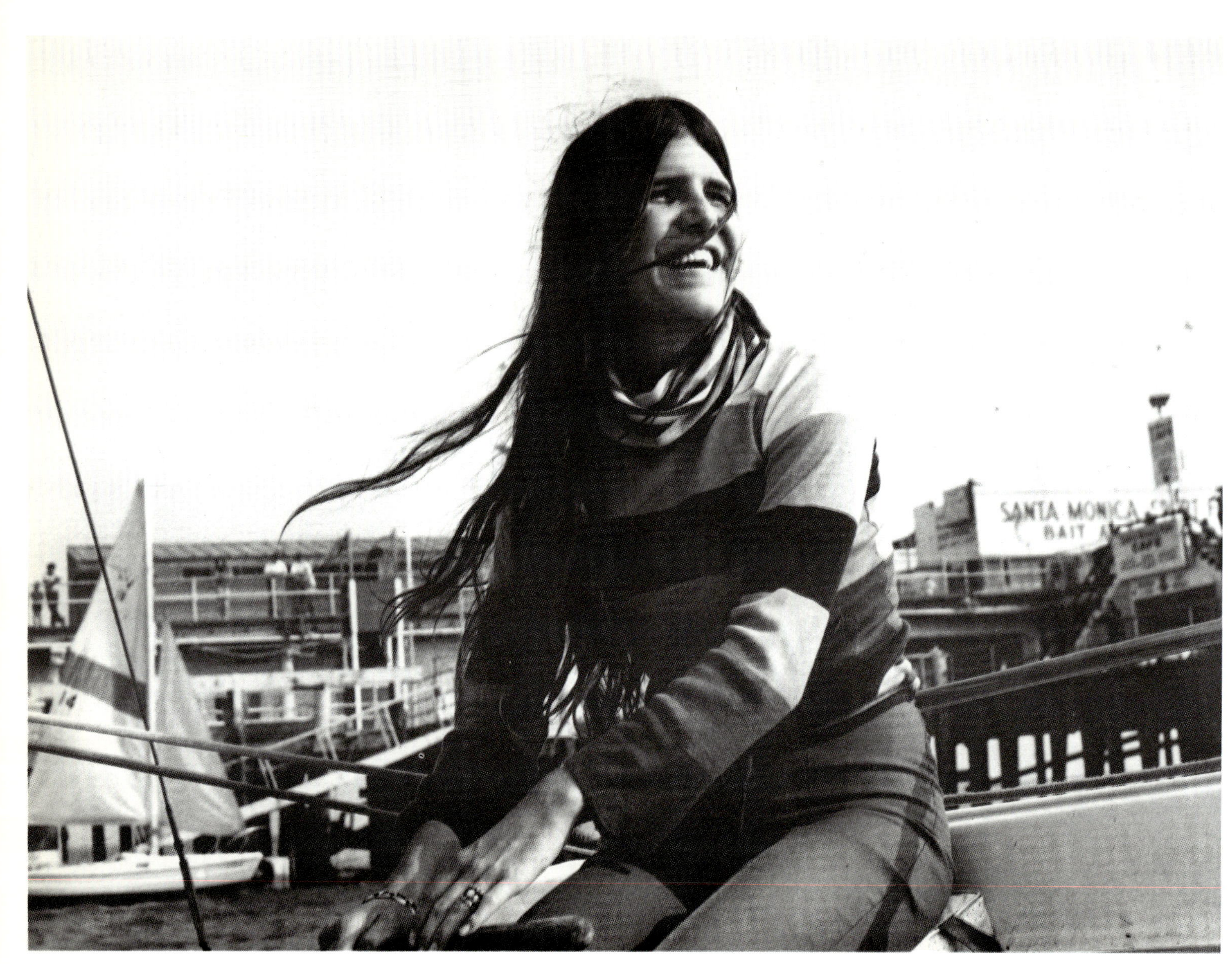

14. Westwind View of the Pier

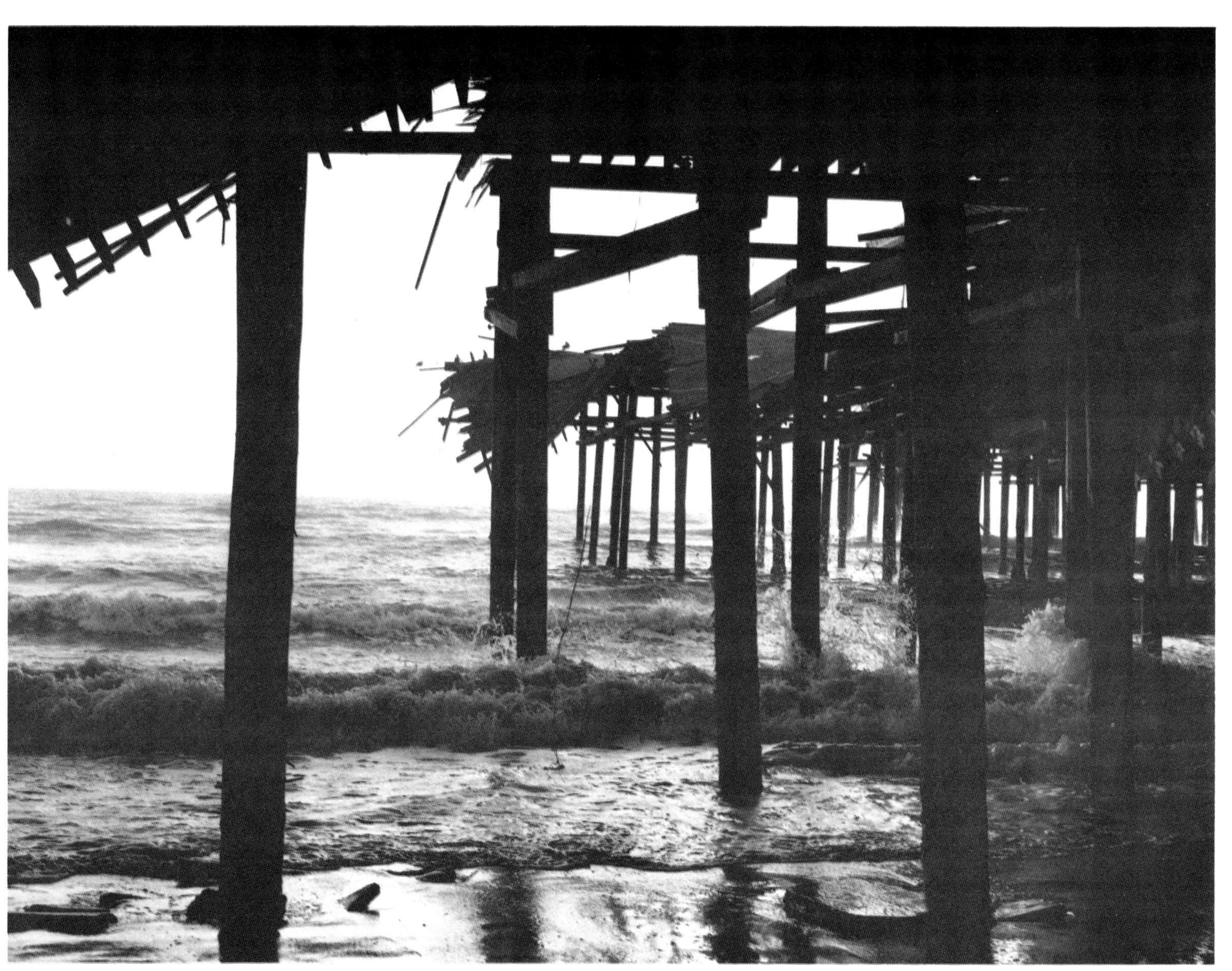

15. Storm Ruins

III: VISITORS

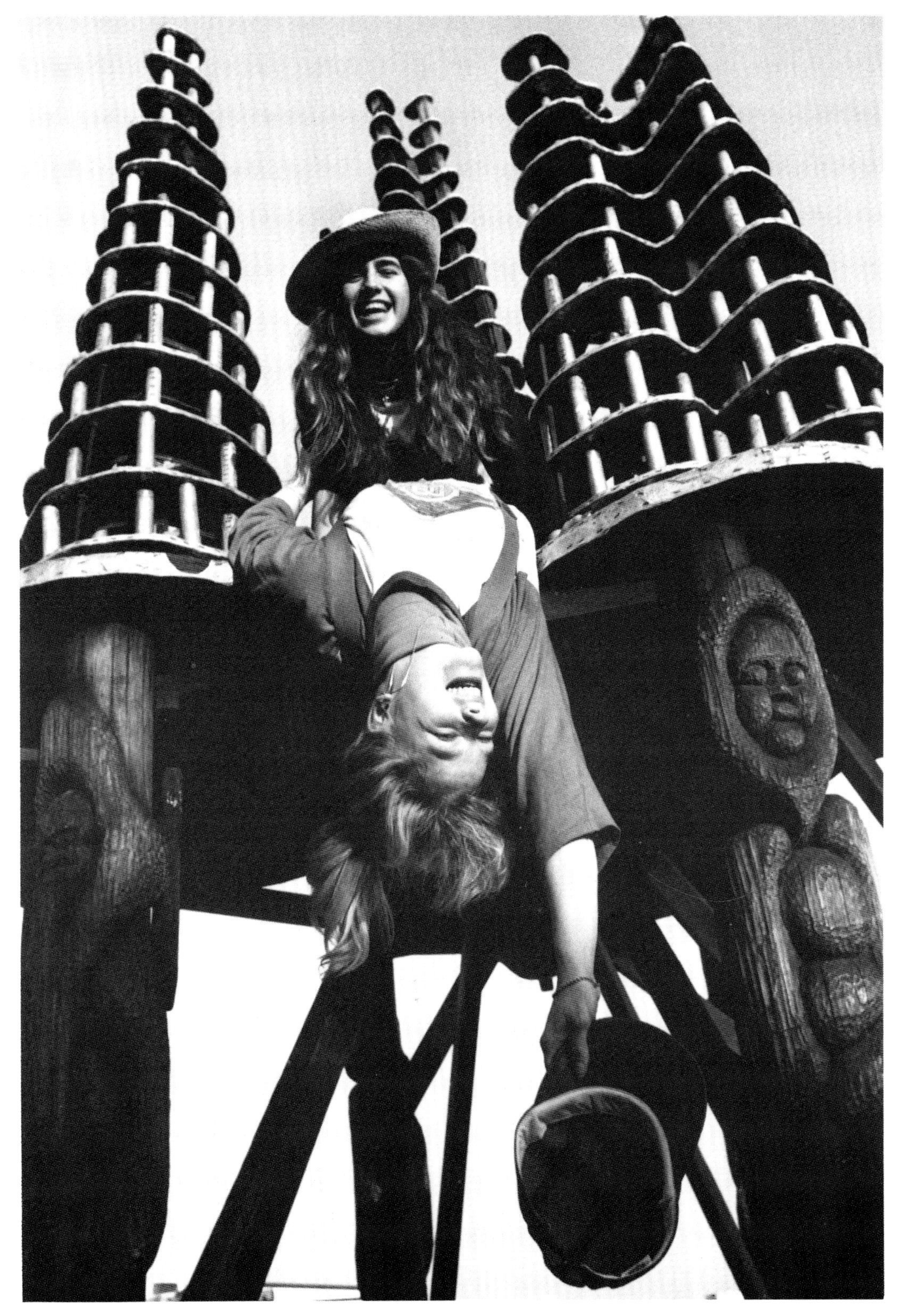

16. Monique and Kathryn

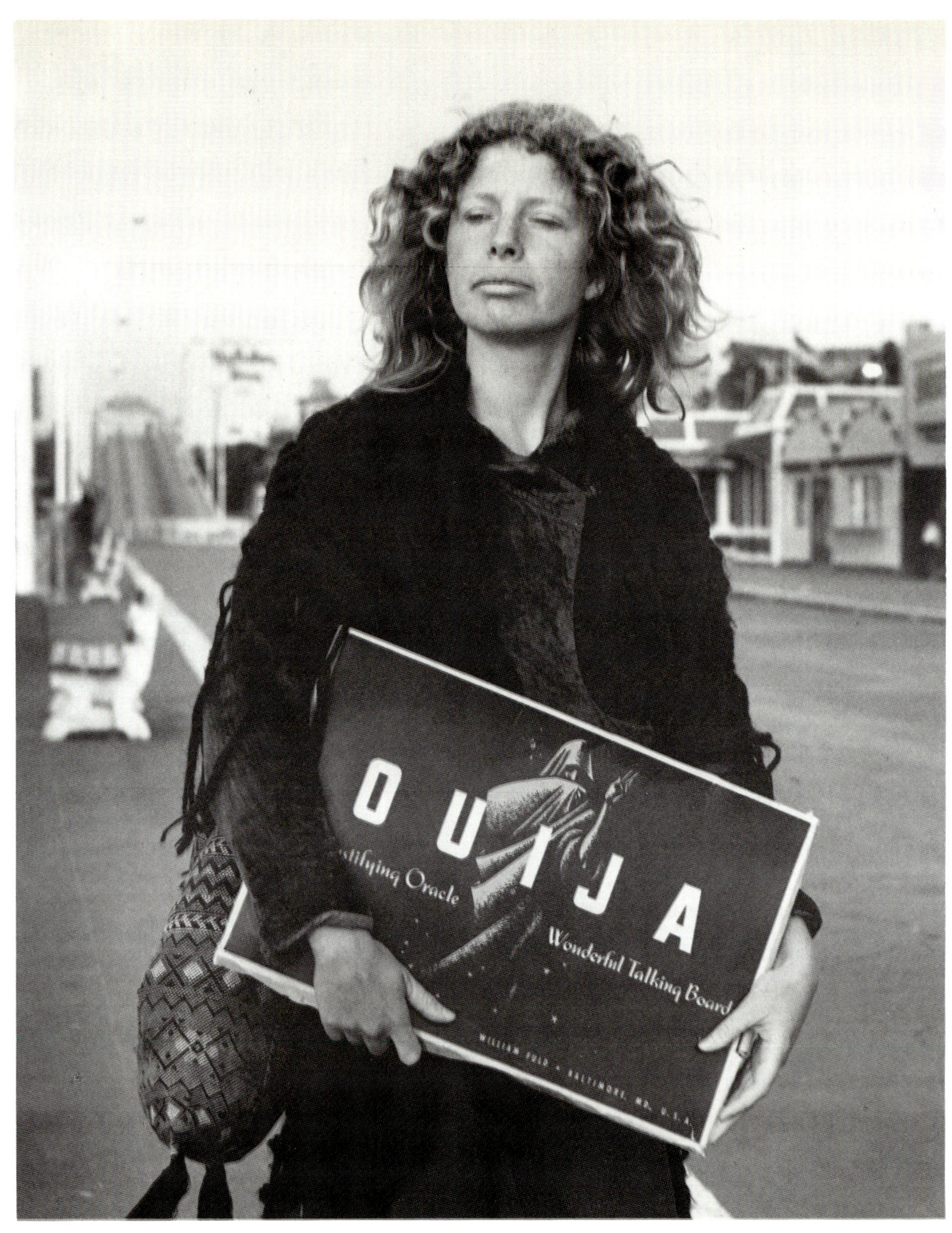

17. Ouija Woman

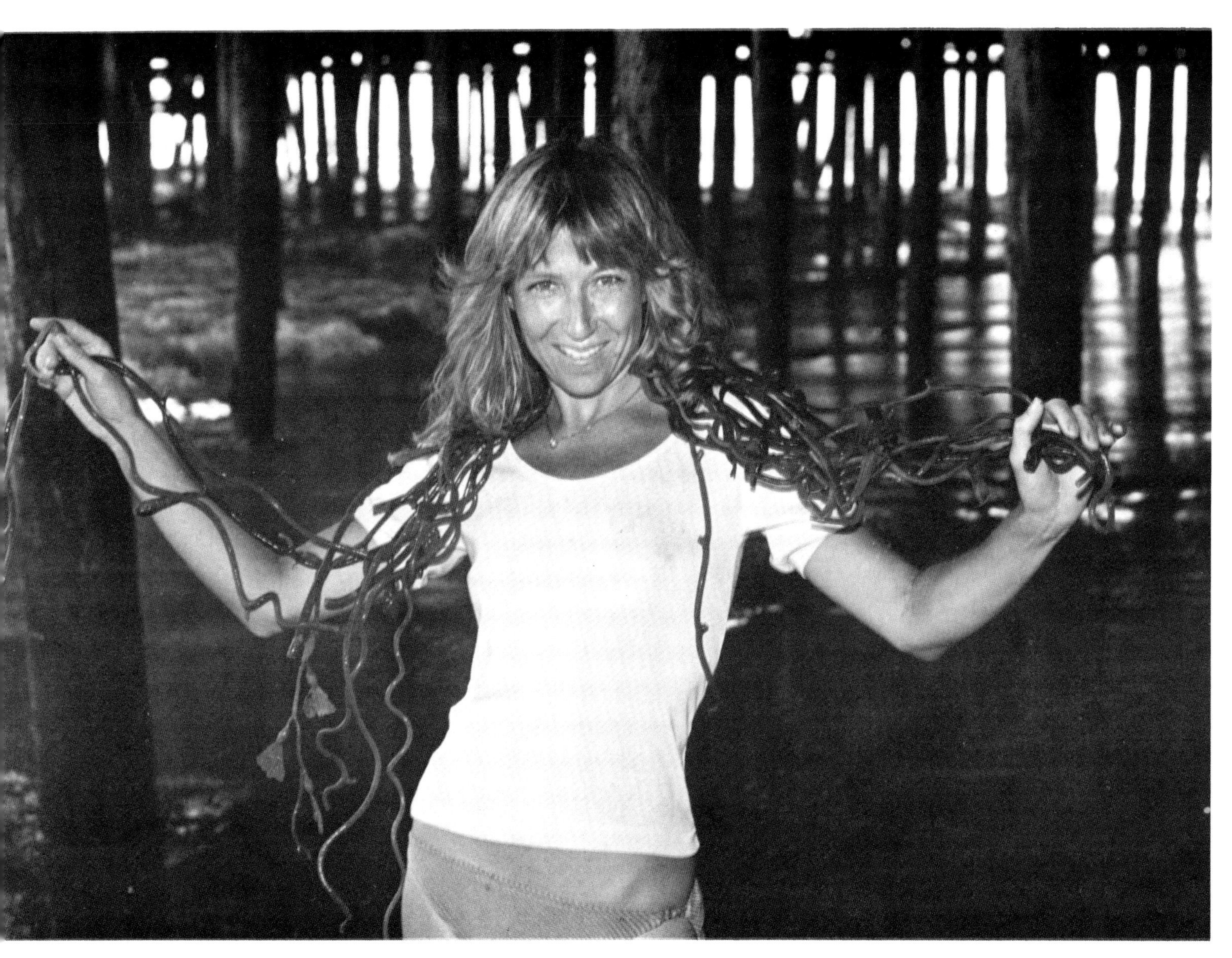

18. Terri, The Seaweed Goddess

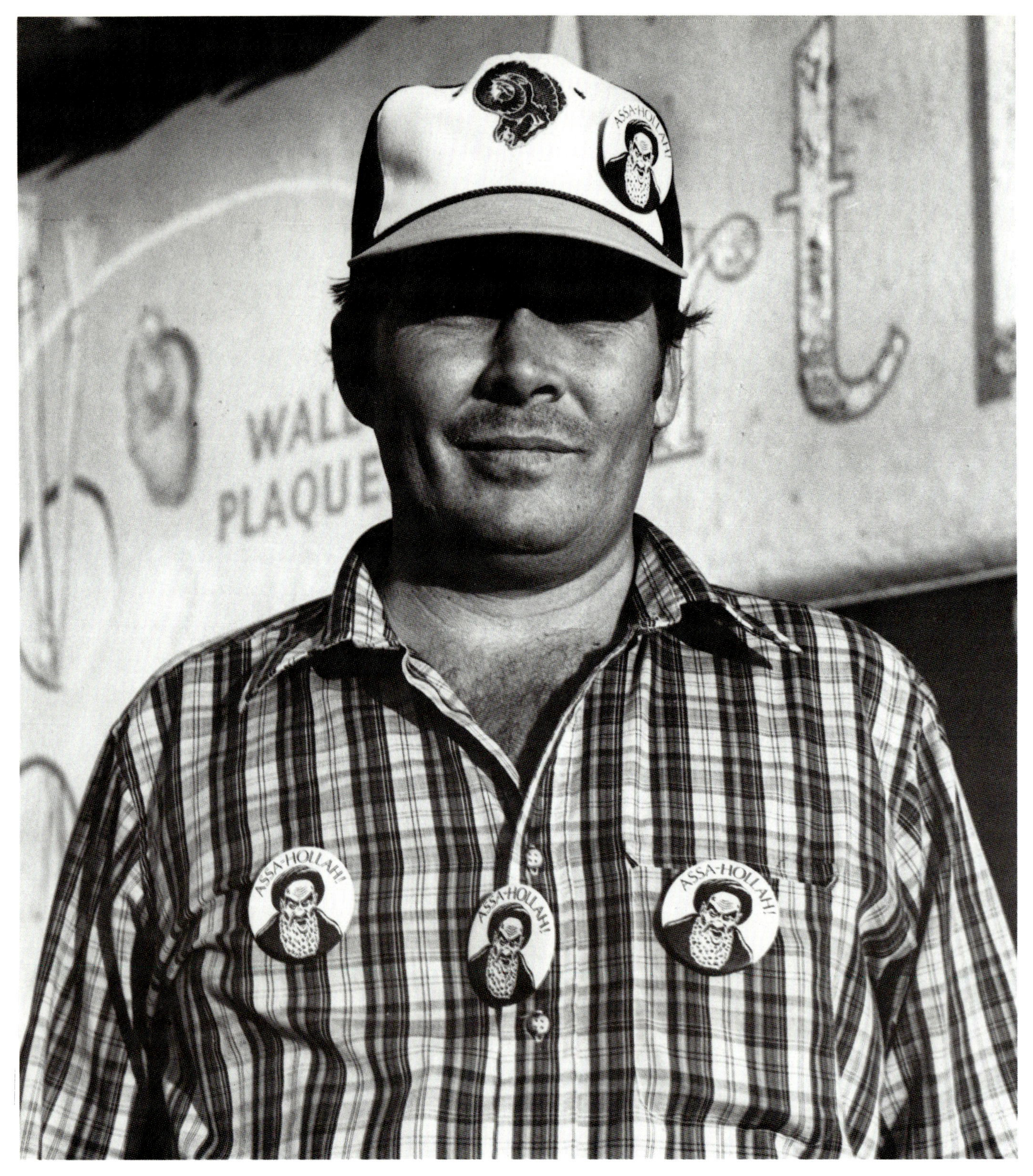

19. Tommy

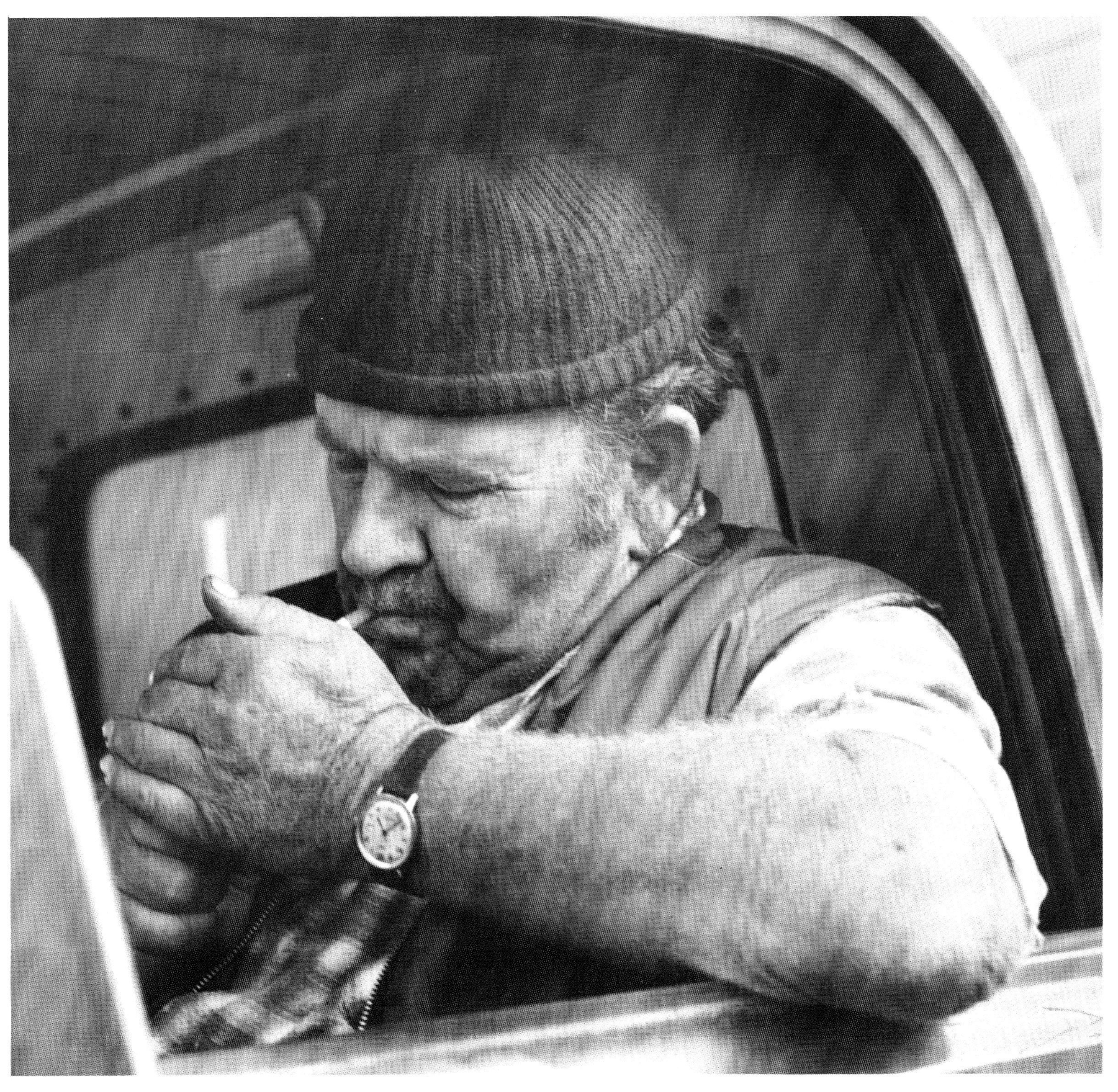

20. Sid, Broken Truck

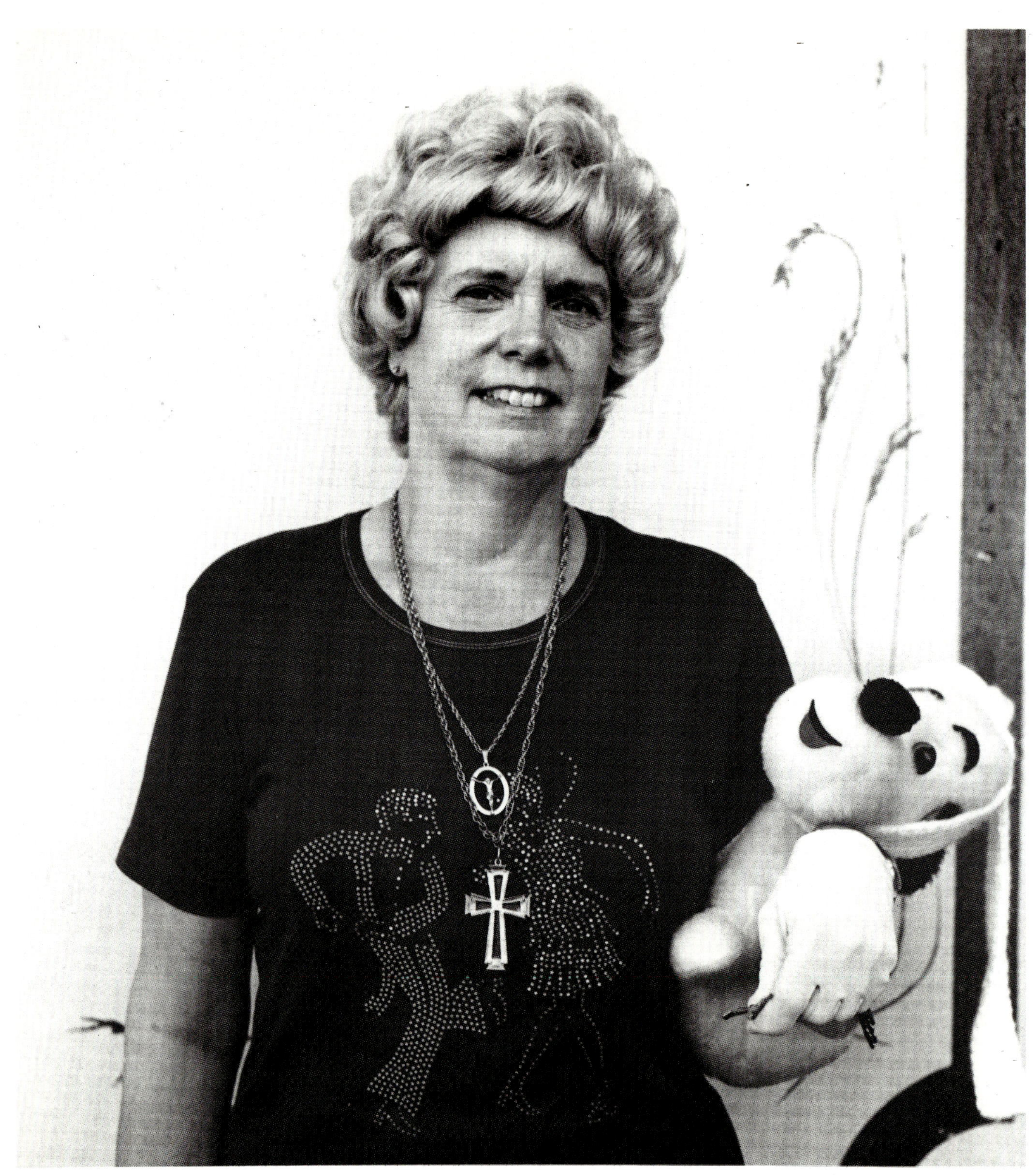

21. Laura and Snoopy

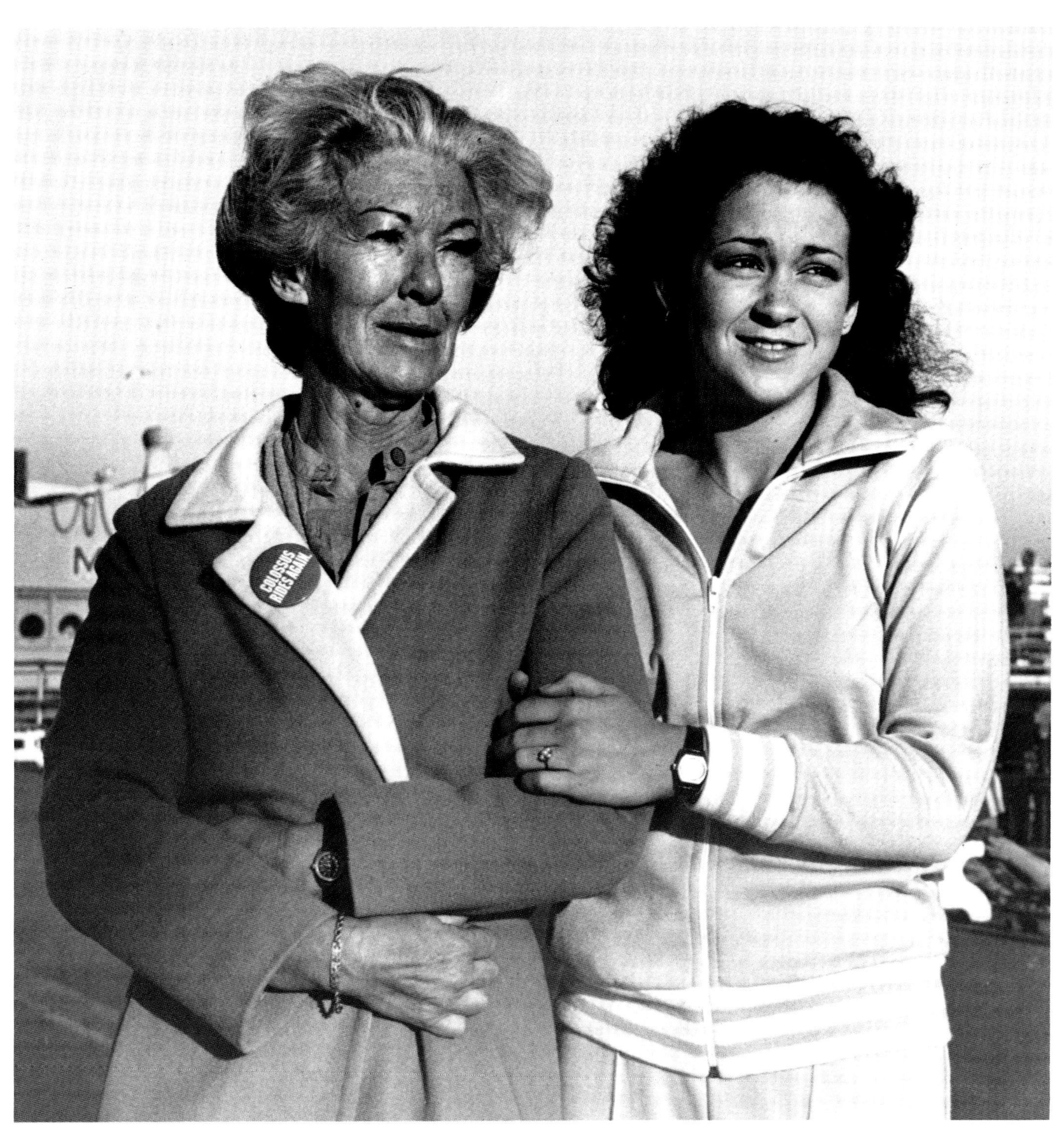

22. Margie and Kim

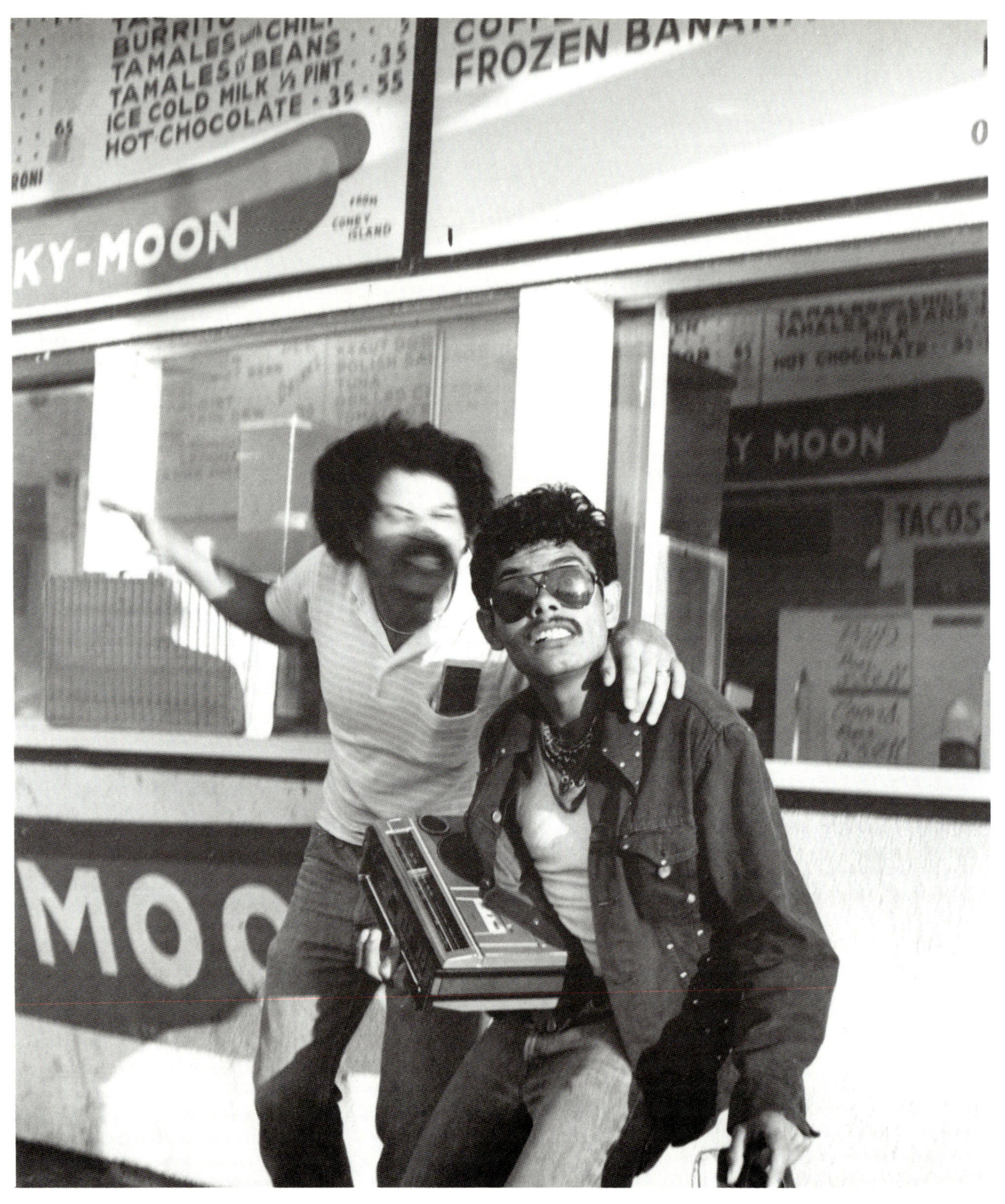

23. George and Raul

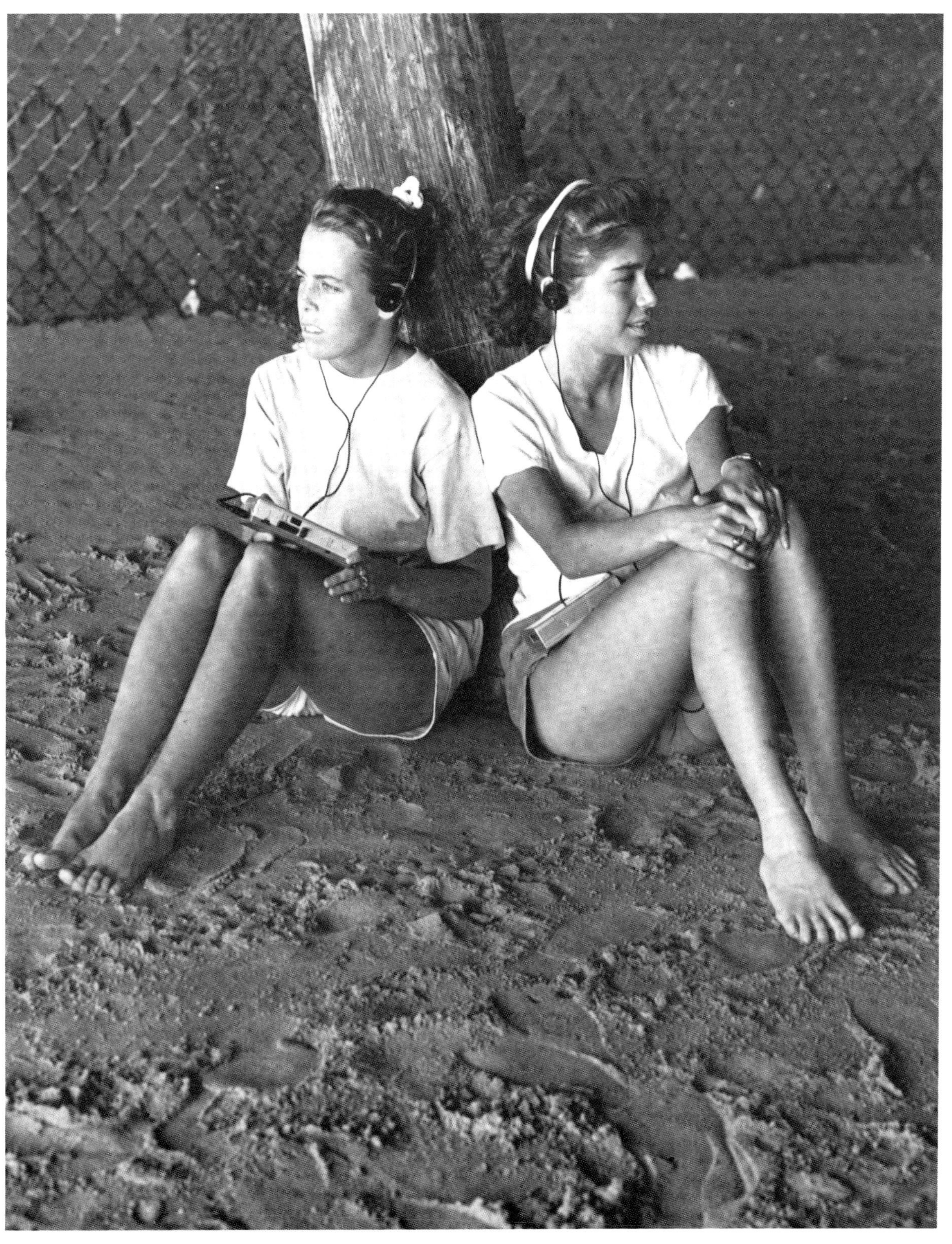

24. Dina and Ellen

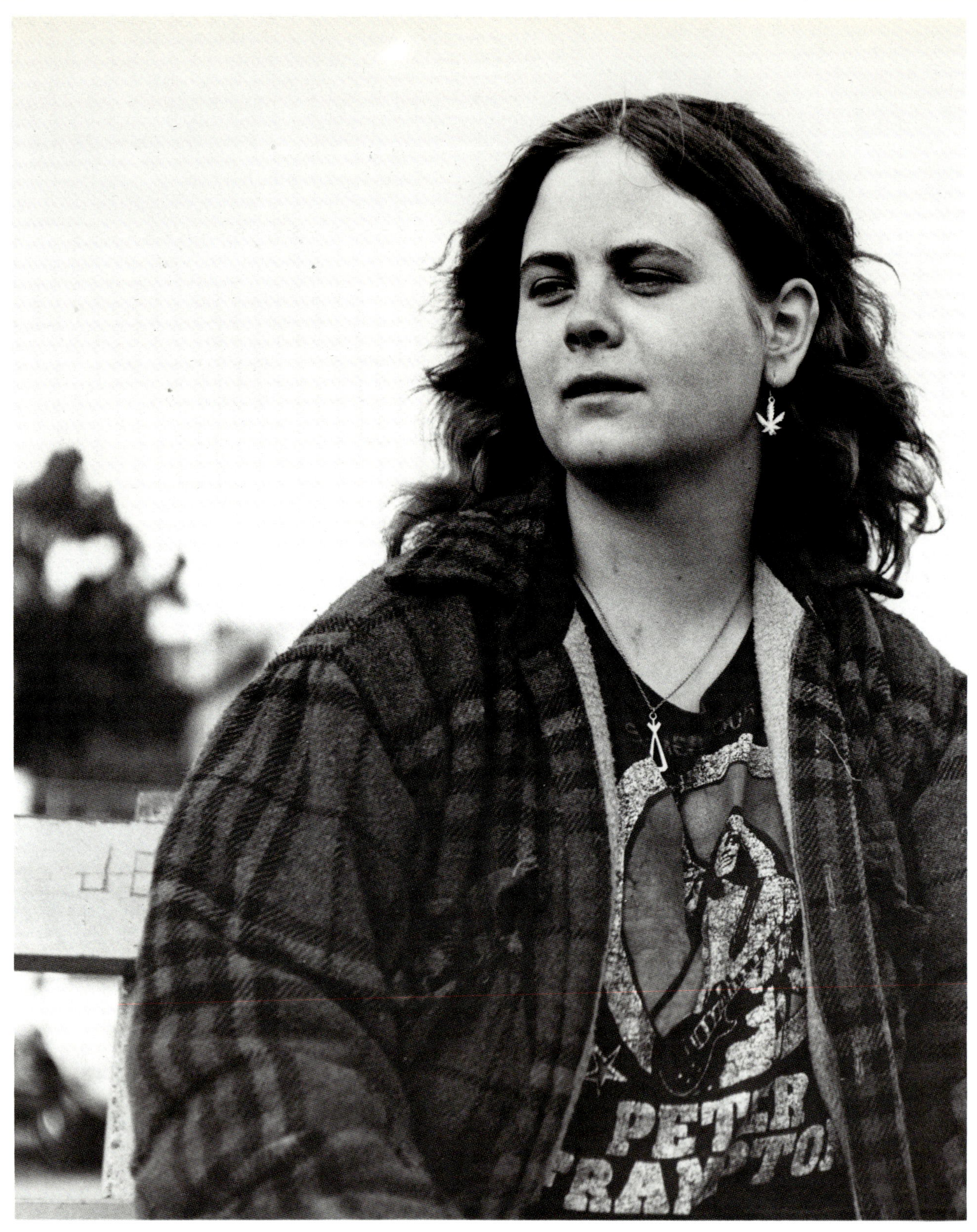

25. Gail

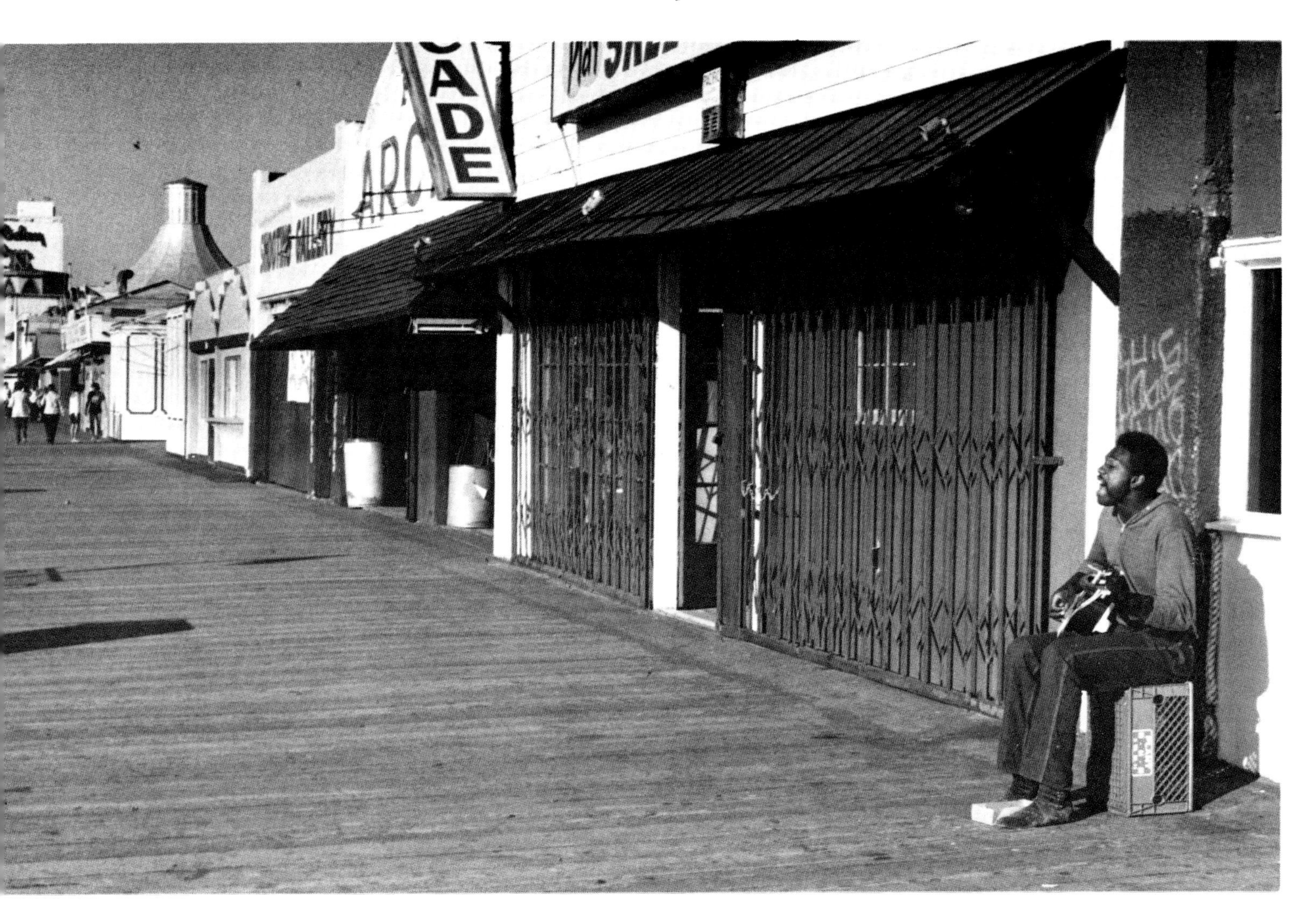

26. Singer from Venice

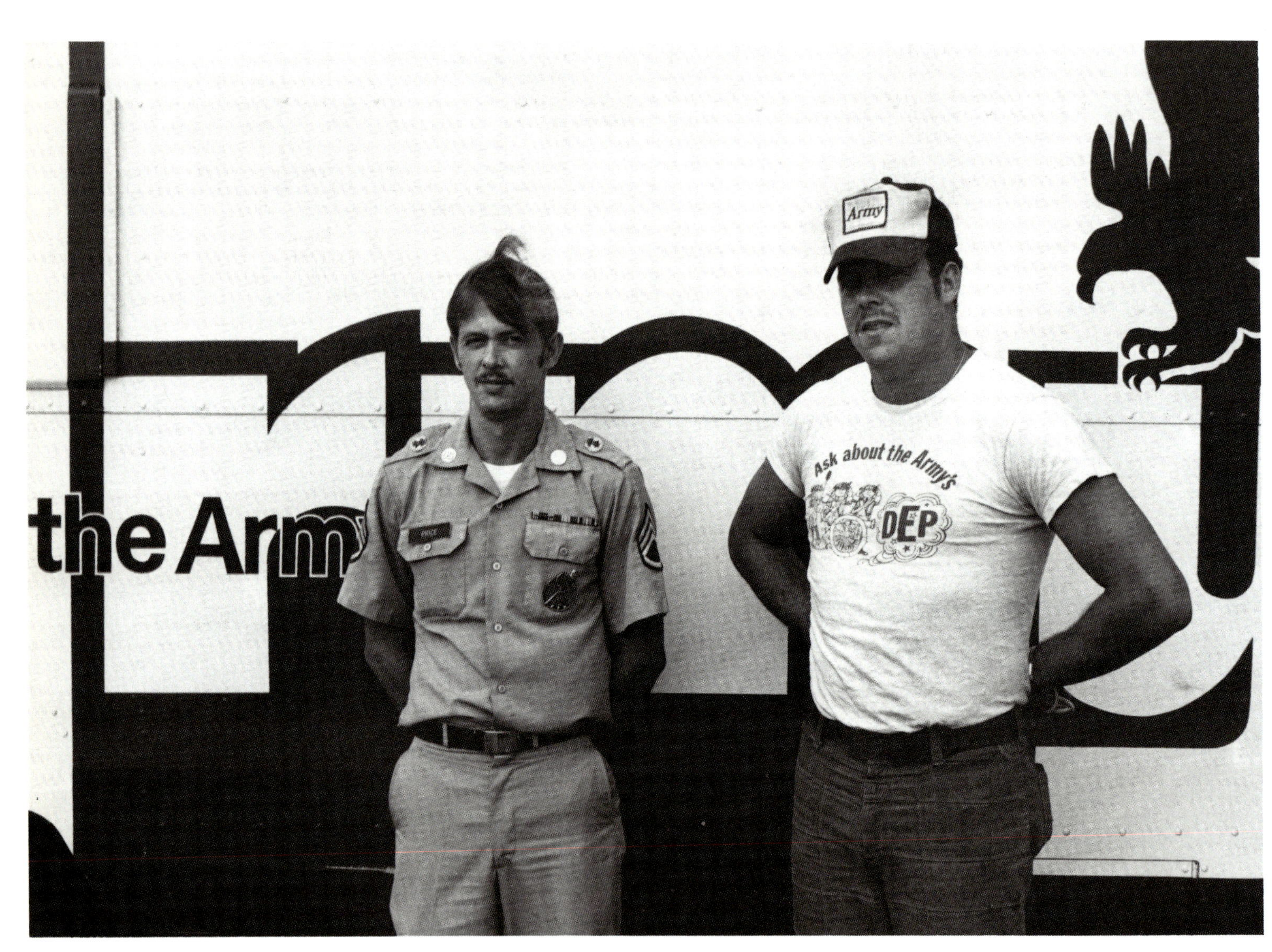

27. Sgts. Rusty Evans and Jay Price

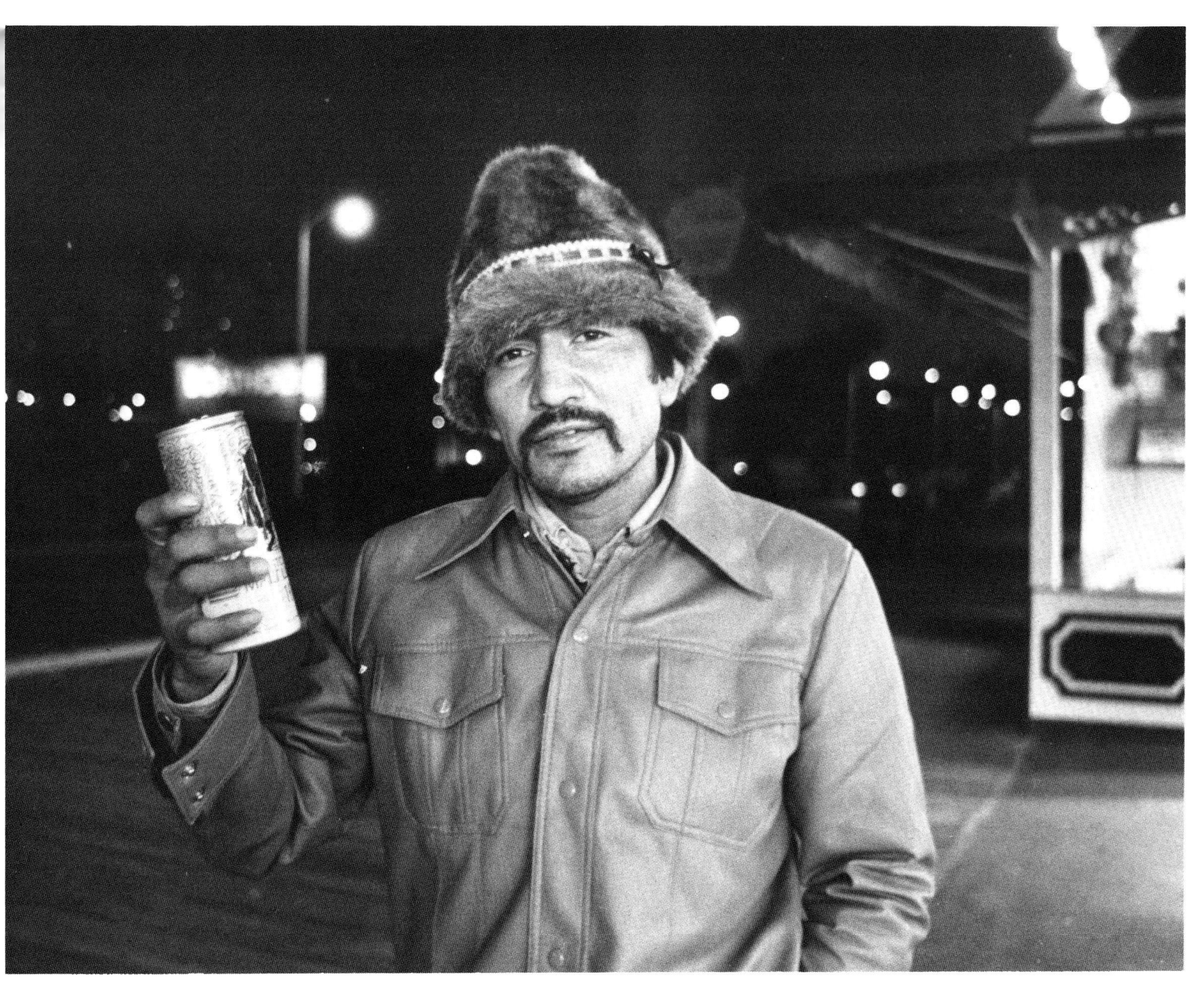

28. Romero

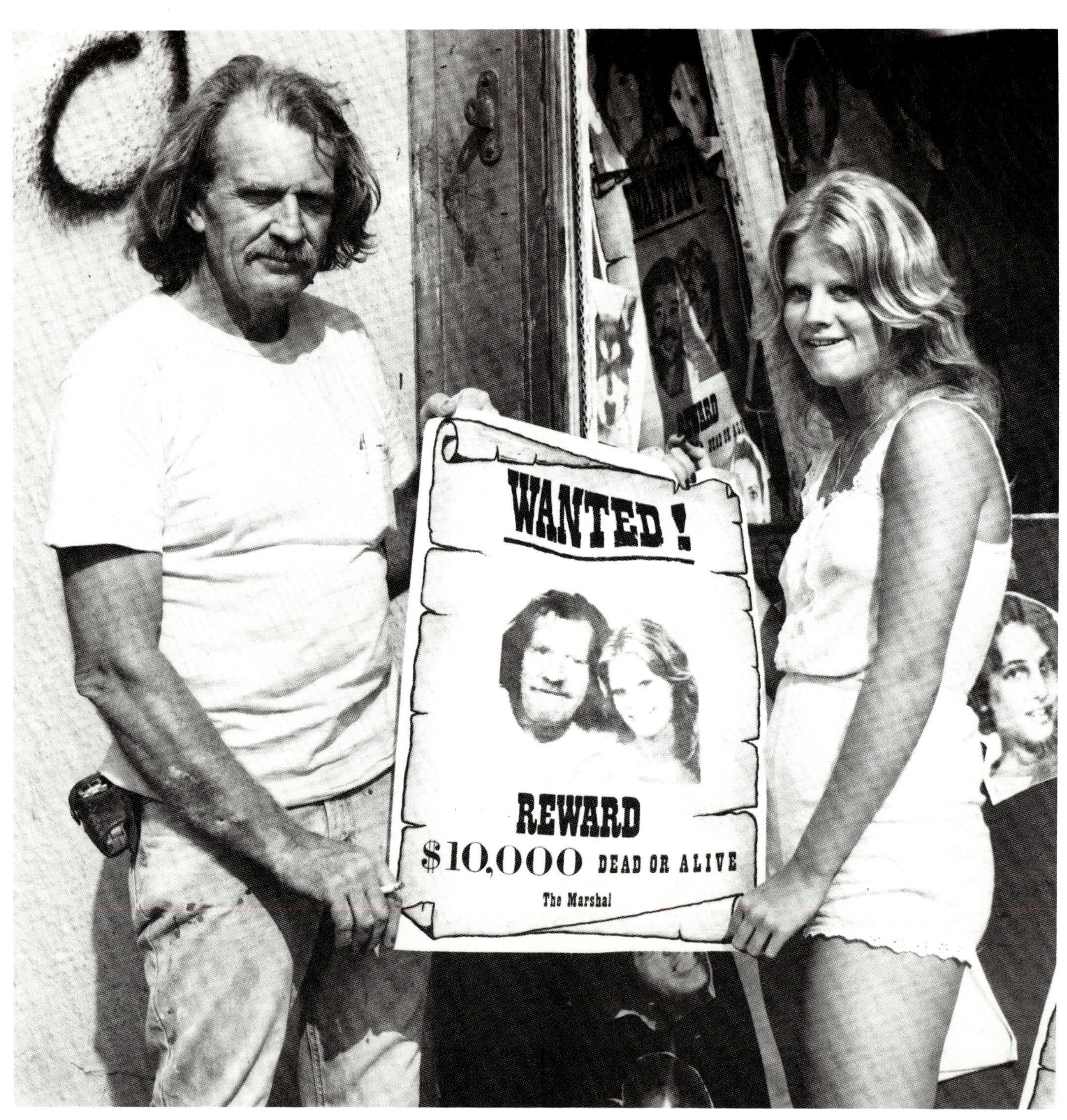

29. Jerimiah and Diana

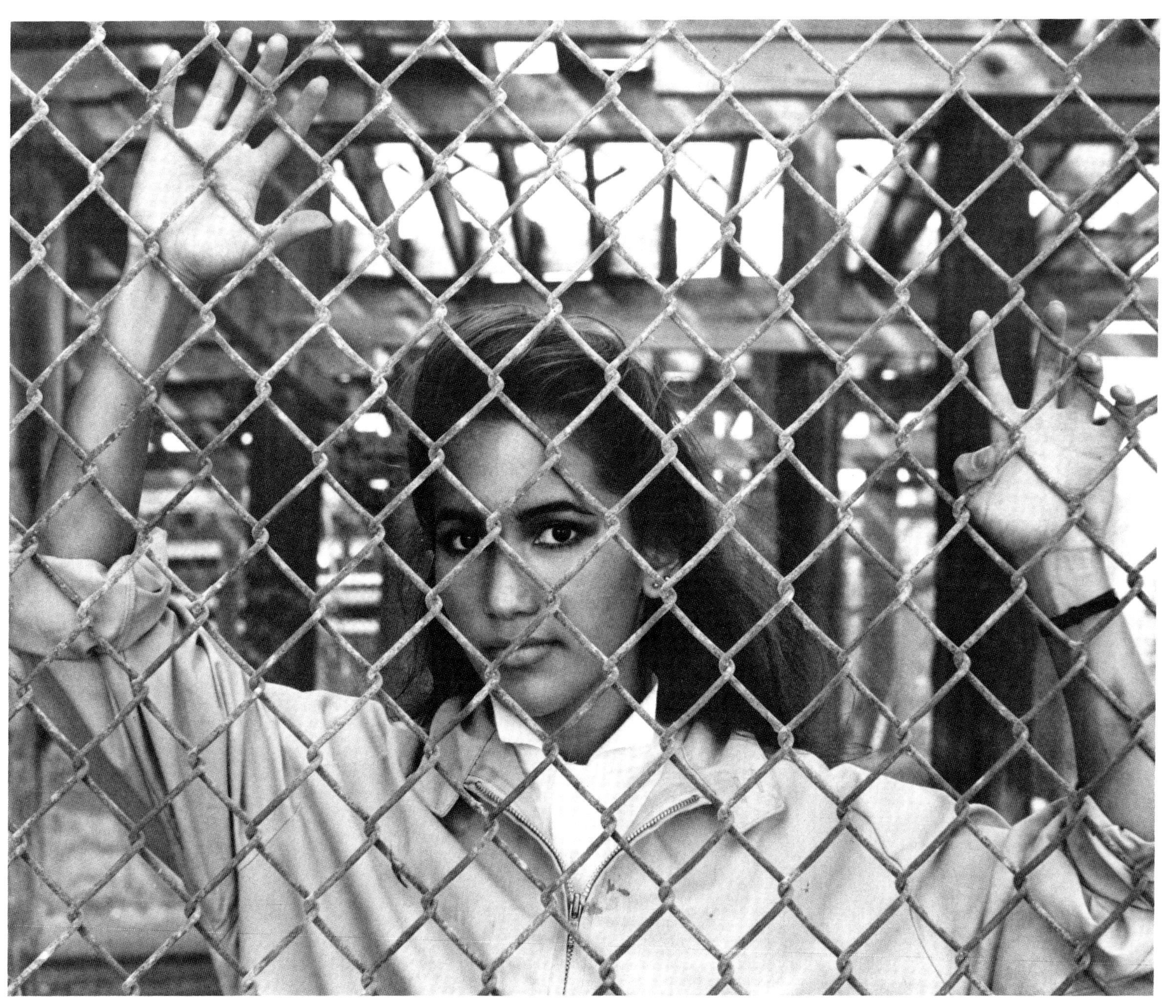

30. Carla

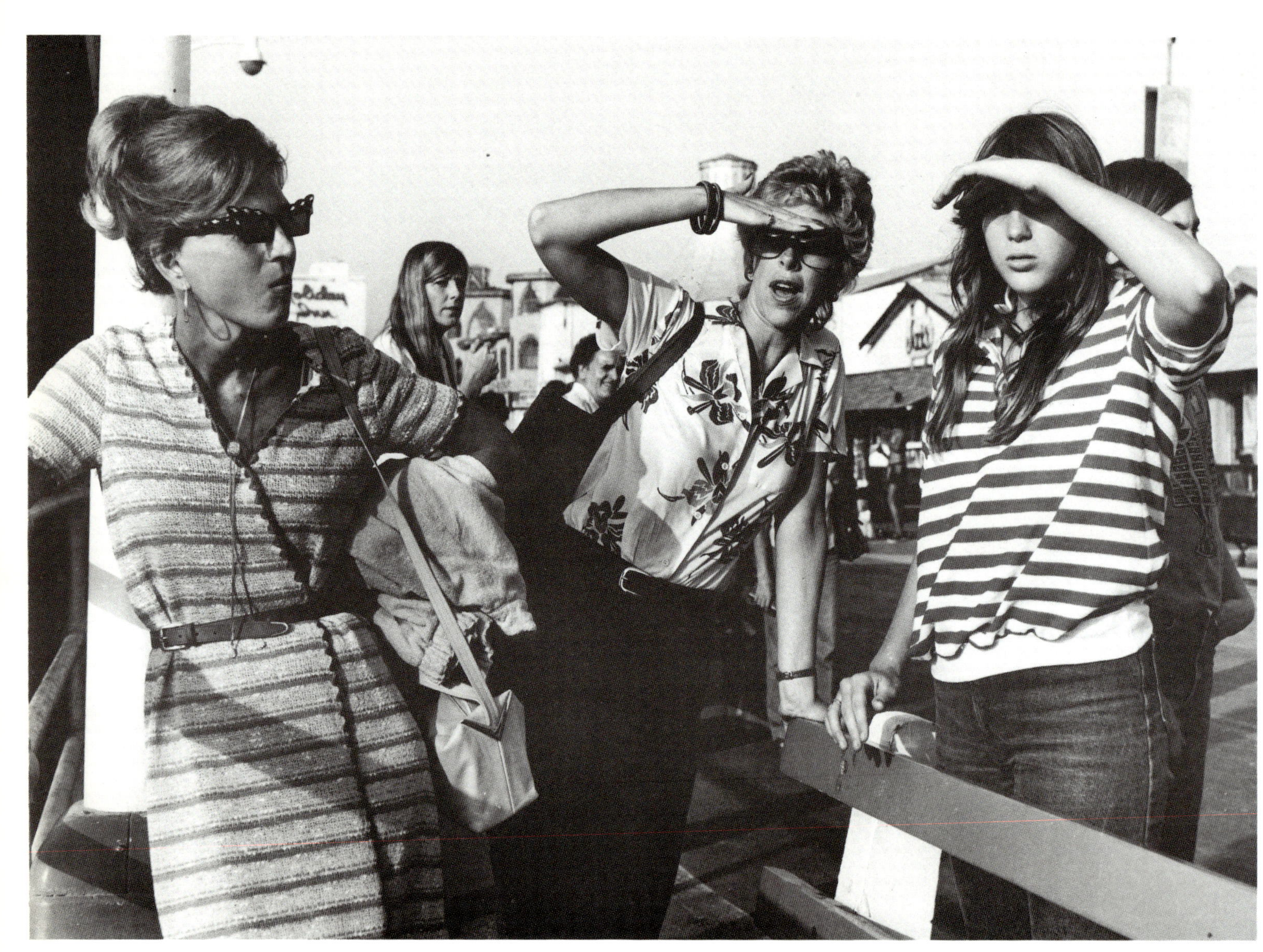

31. Mid-Winter Tourists

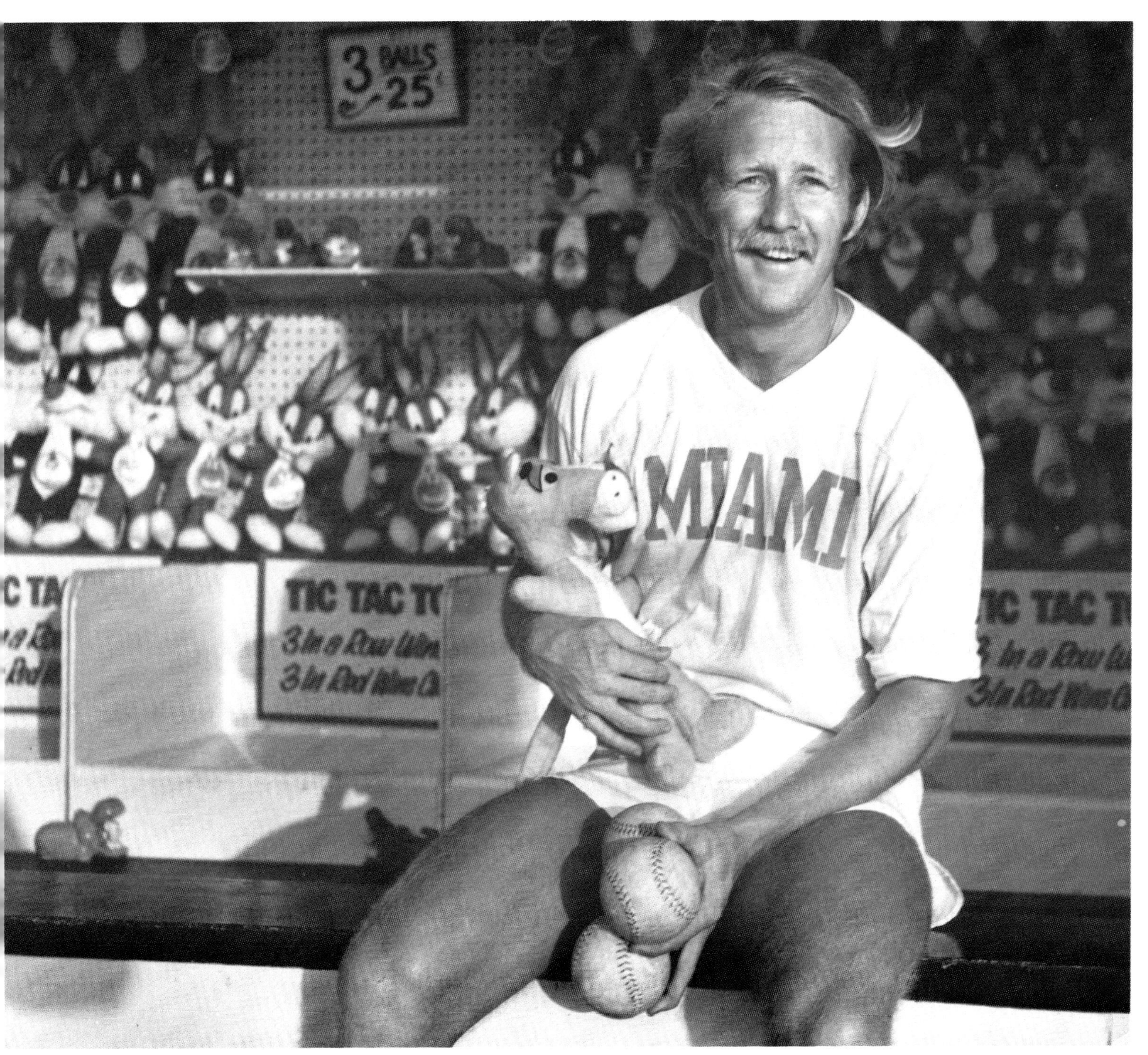

32. Utzer

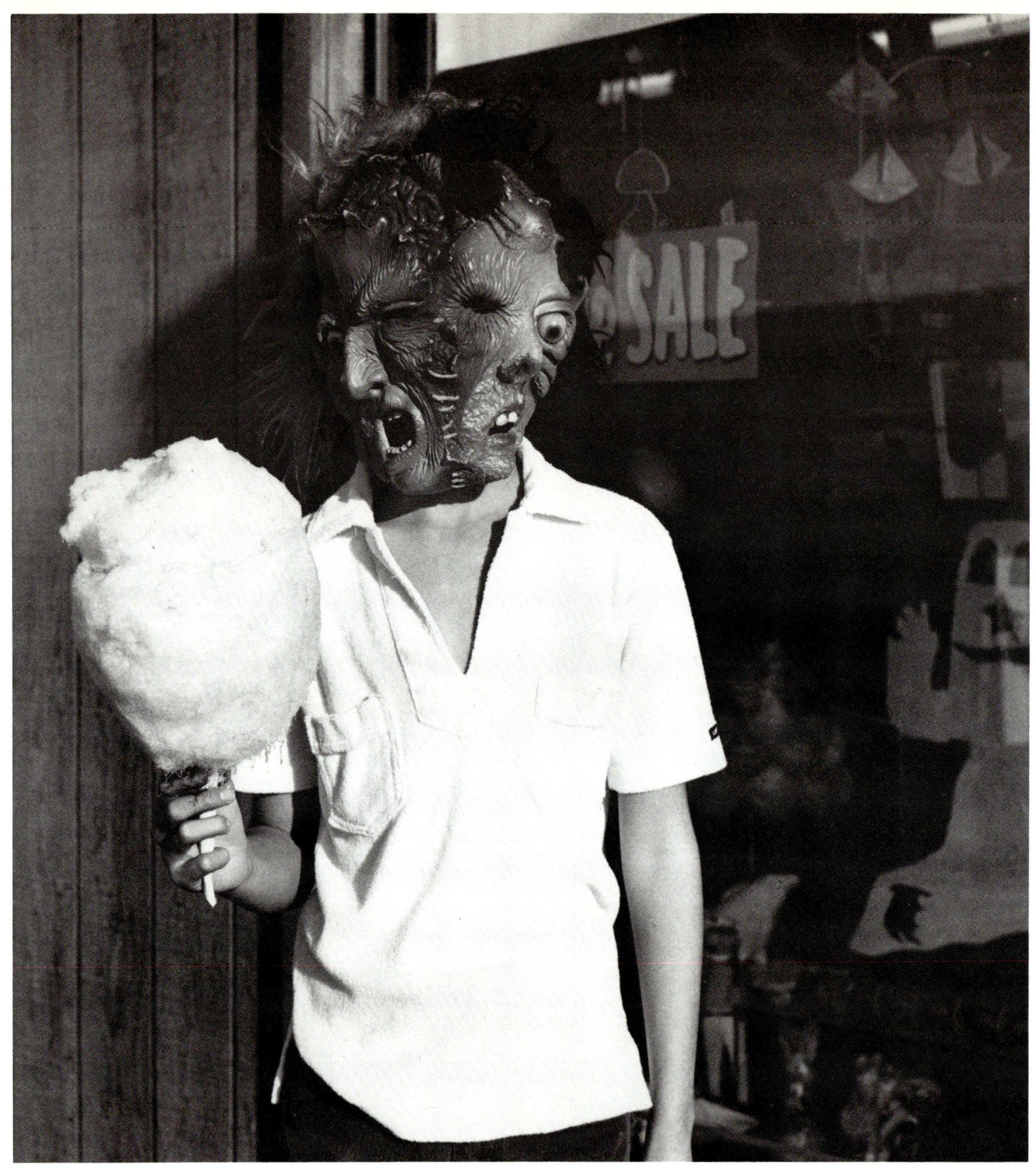

33. Chris

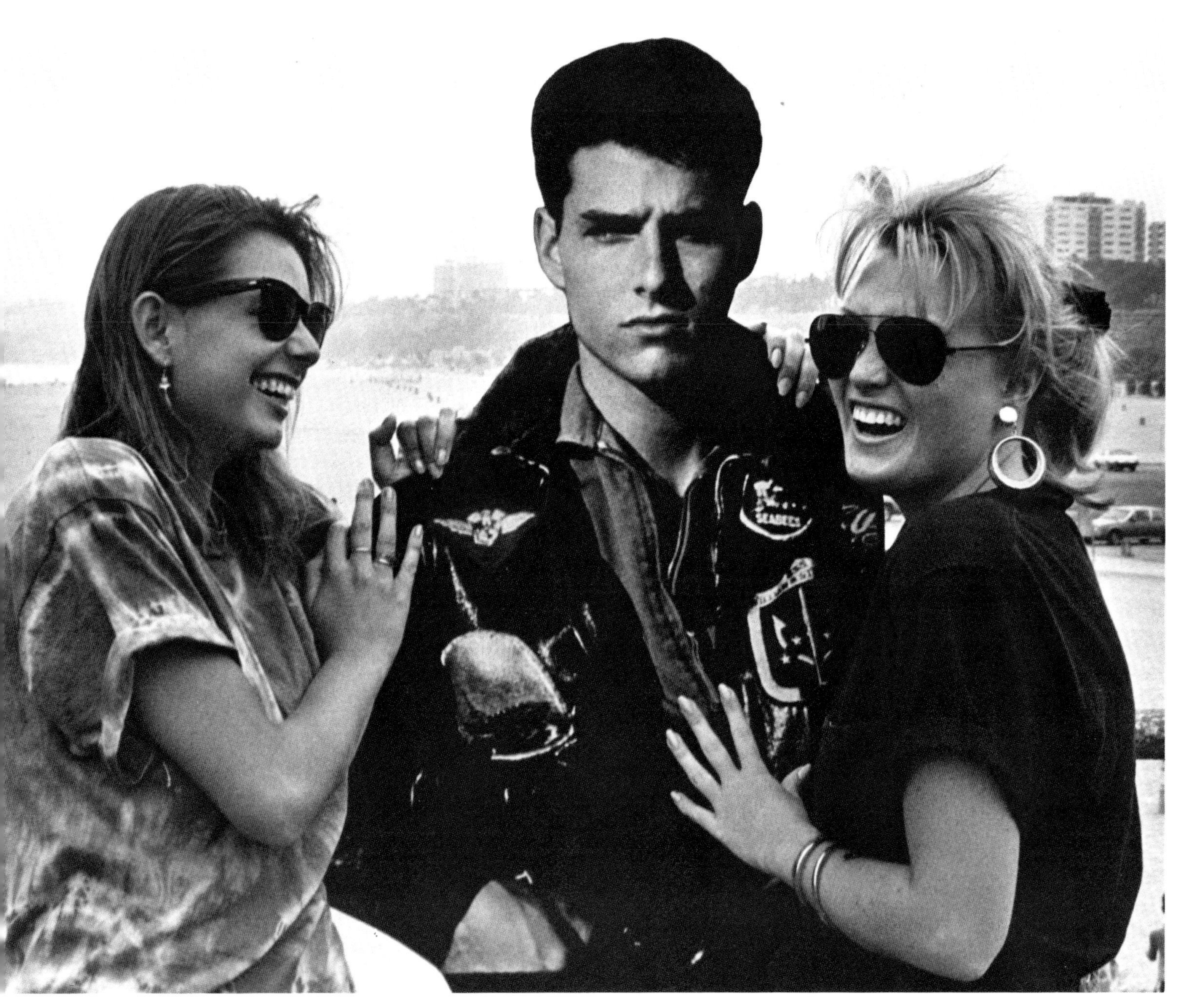

34. Carol, Caterina and "Tom Cruise"

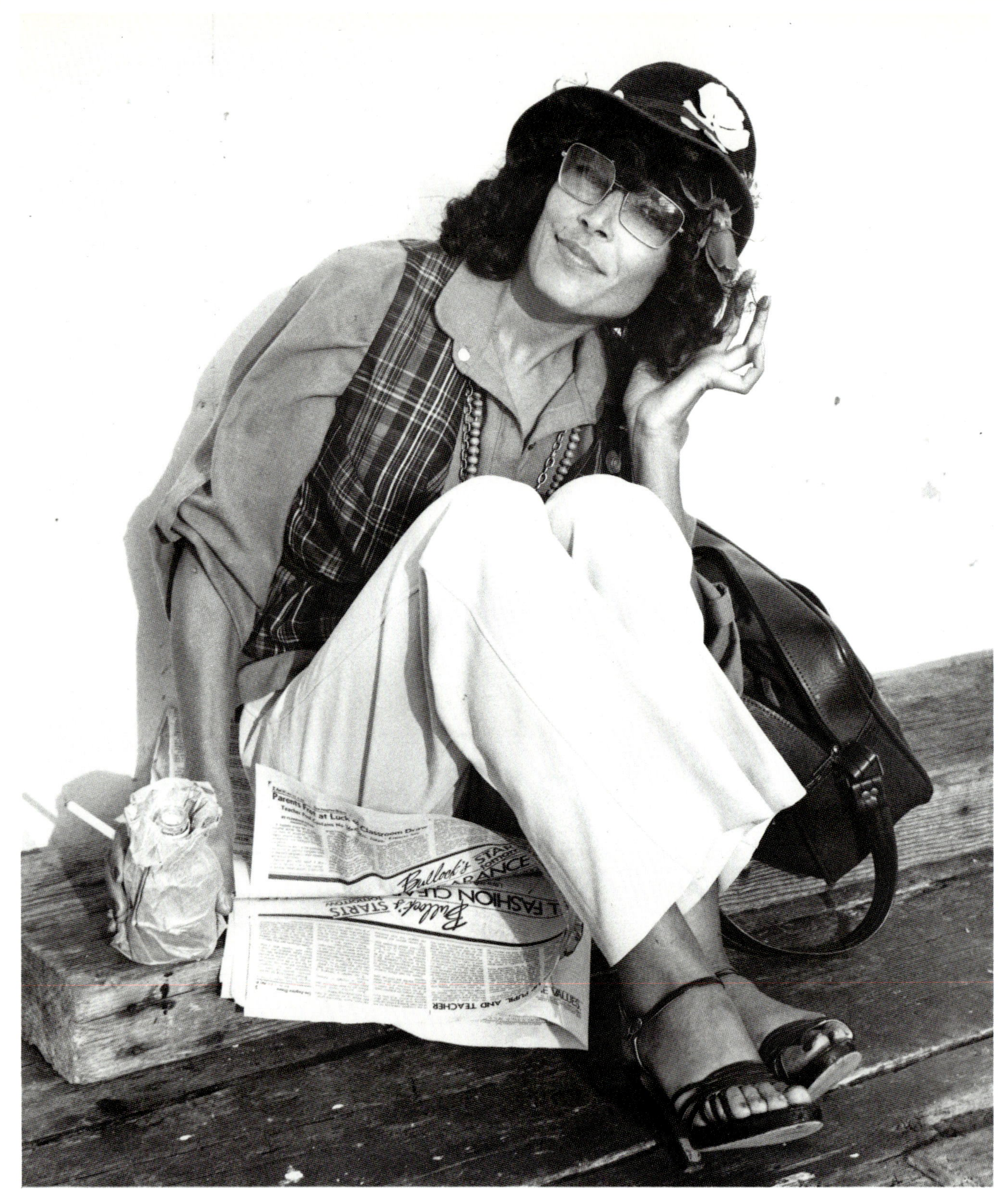

35. Lupe

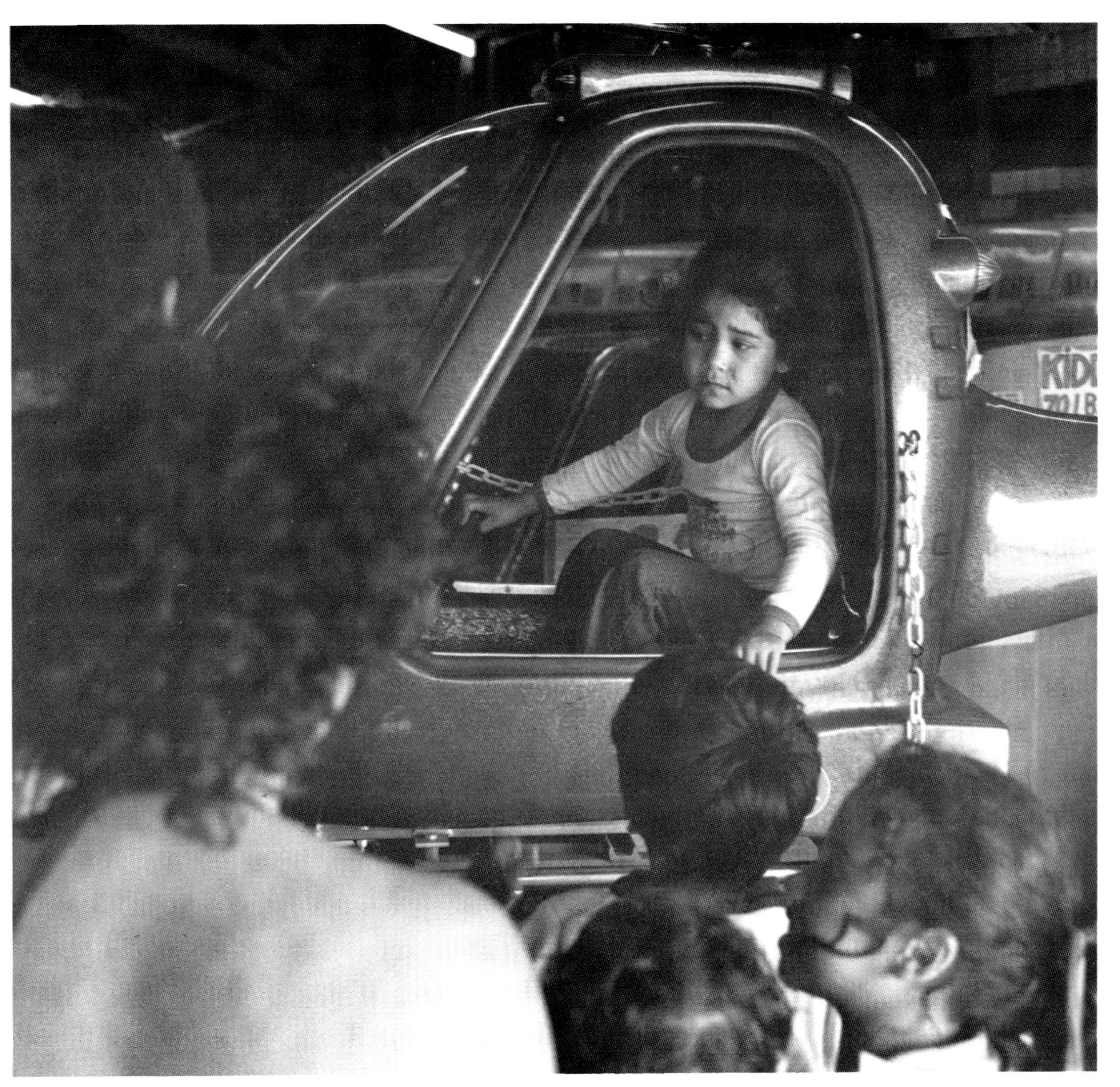

36. Helicopter Ride, Arcade

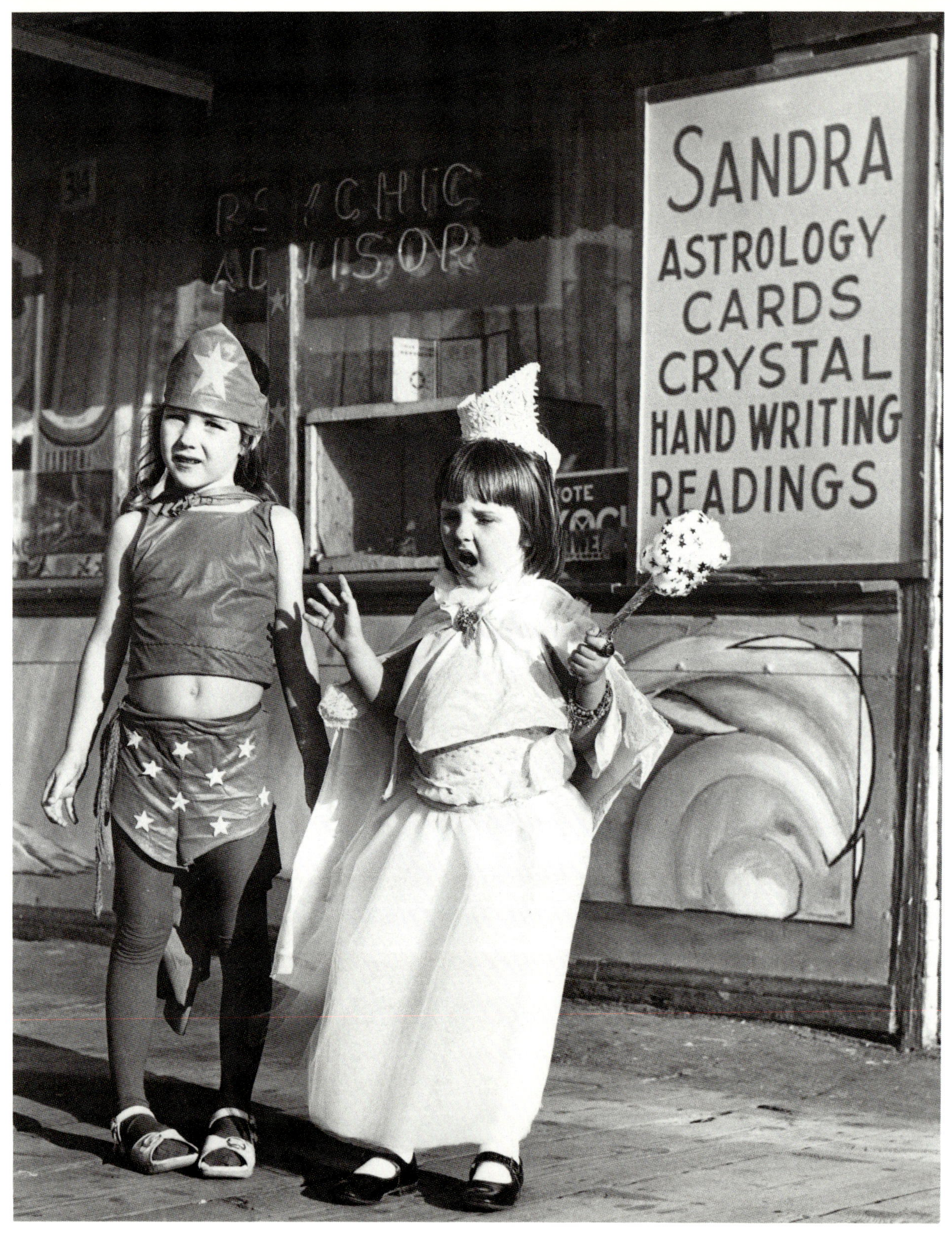

37. Adrian and Elizabeth

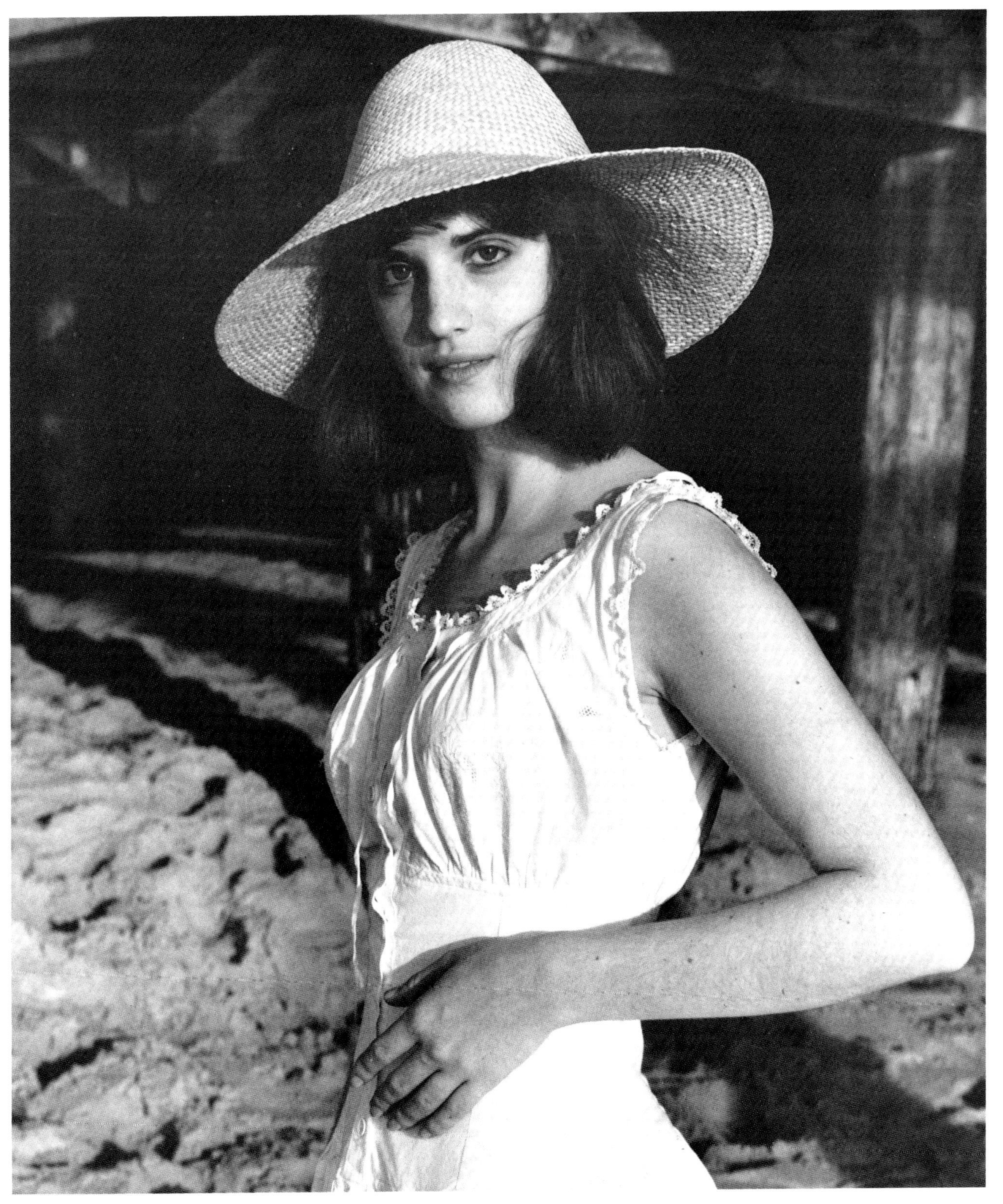

38. Jeannie

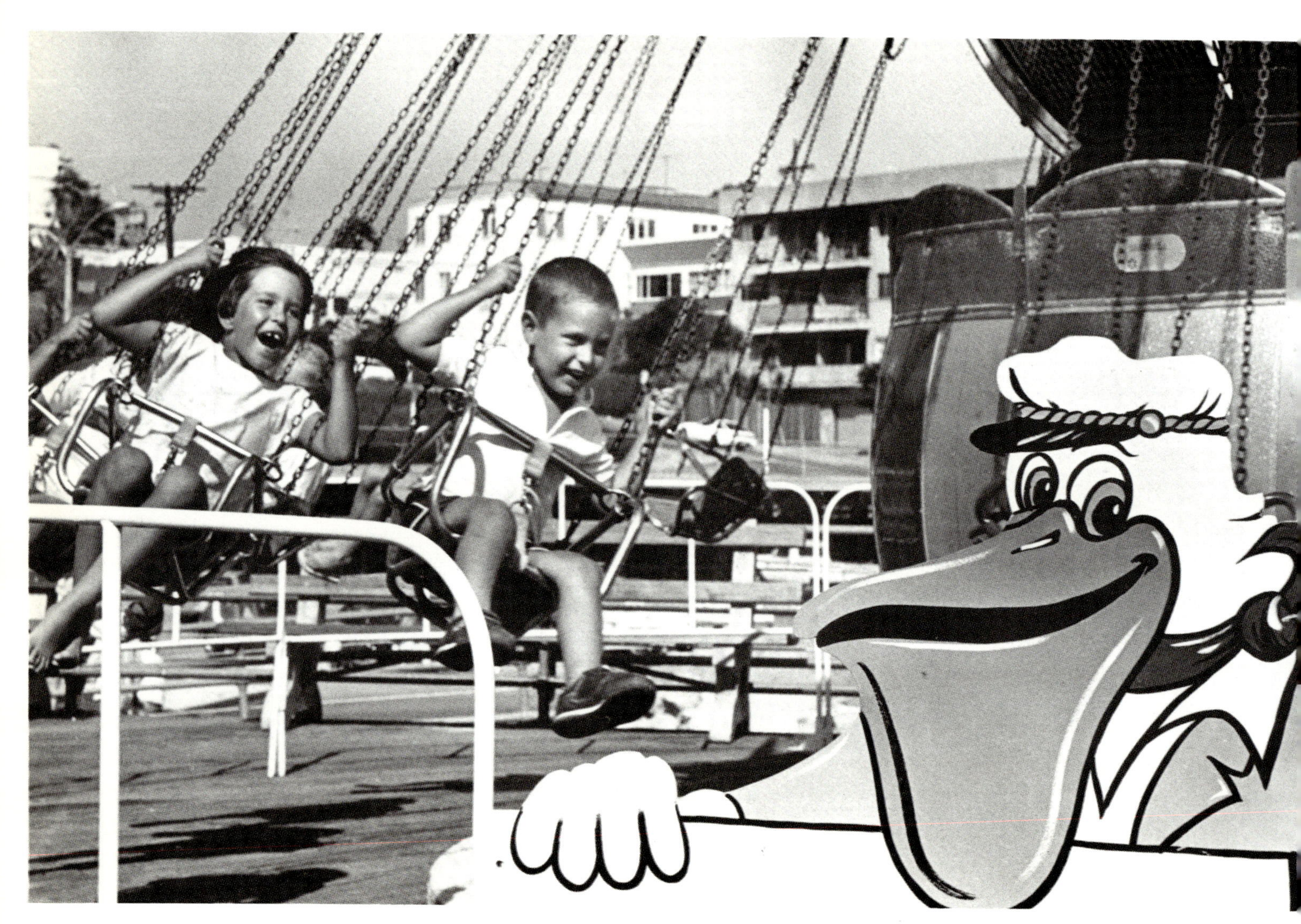

39. Fun Zone

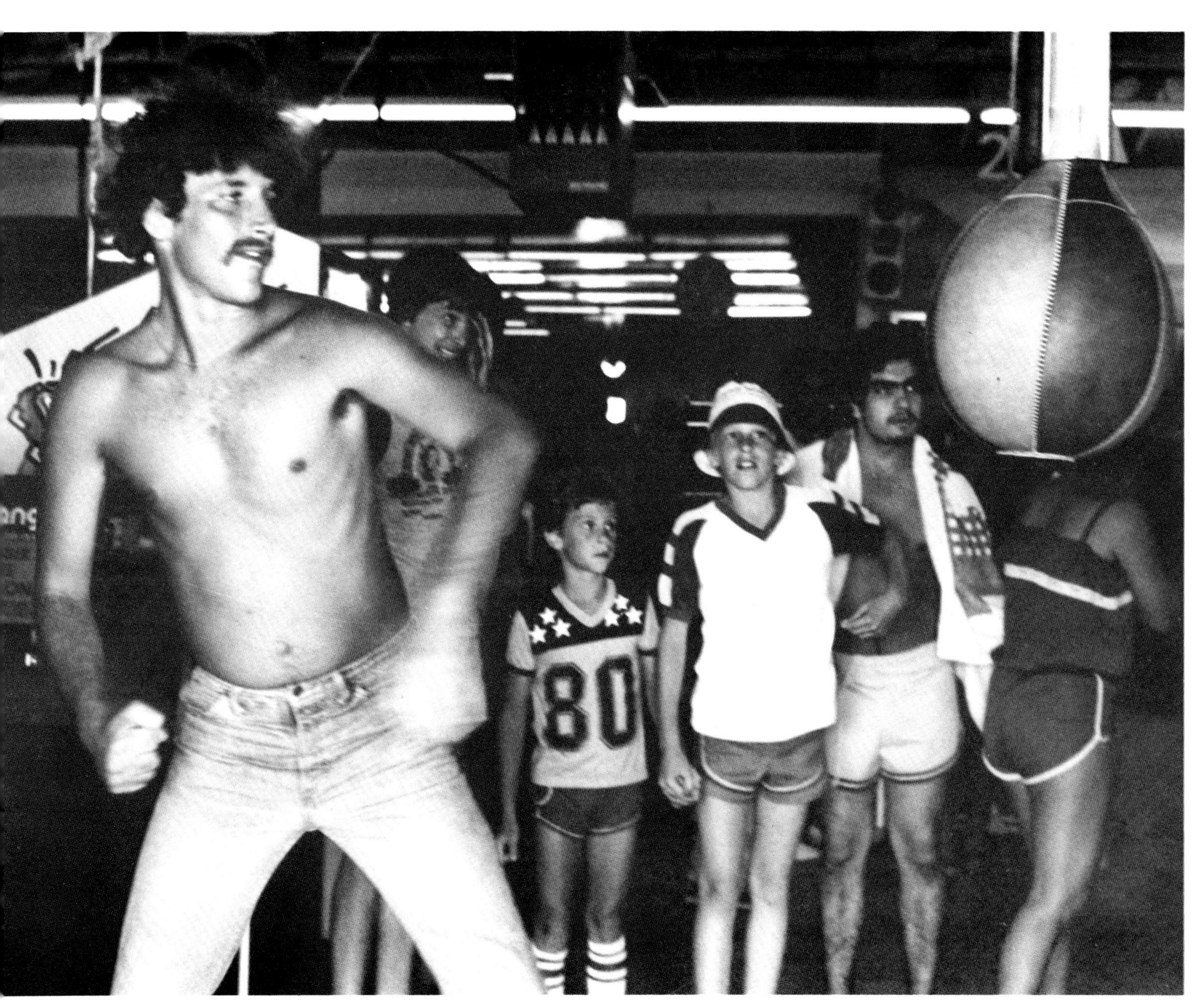

40. The Punching Bag

IV: REGULARS

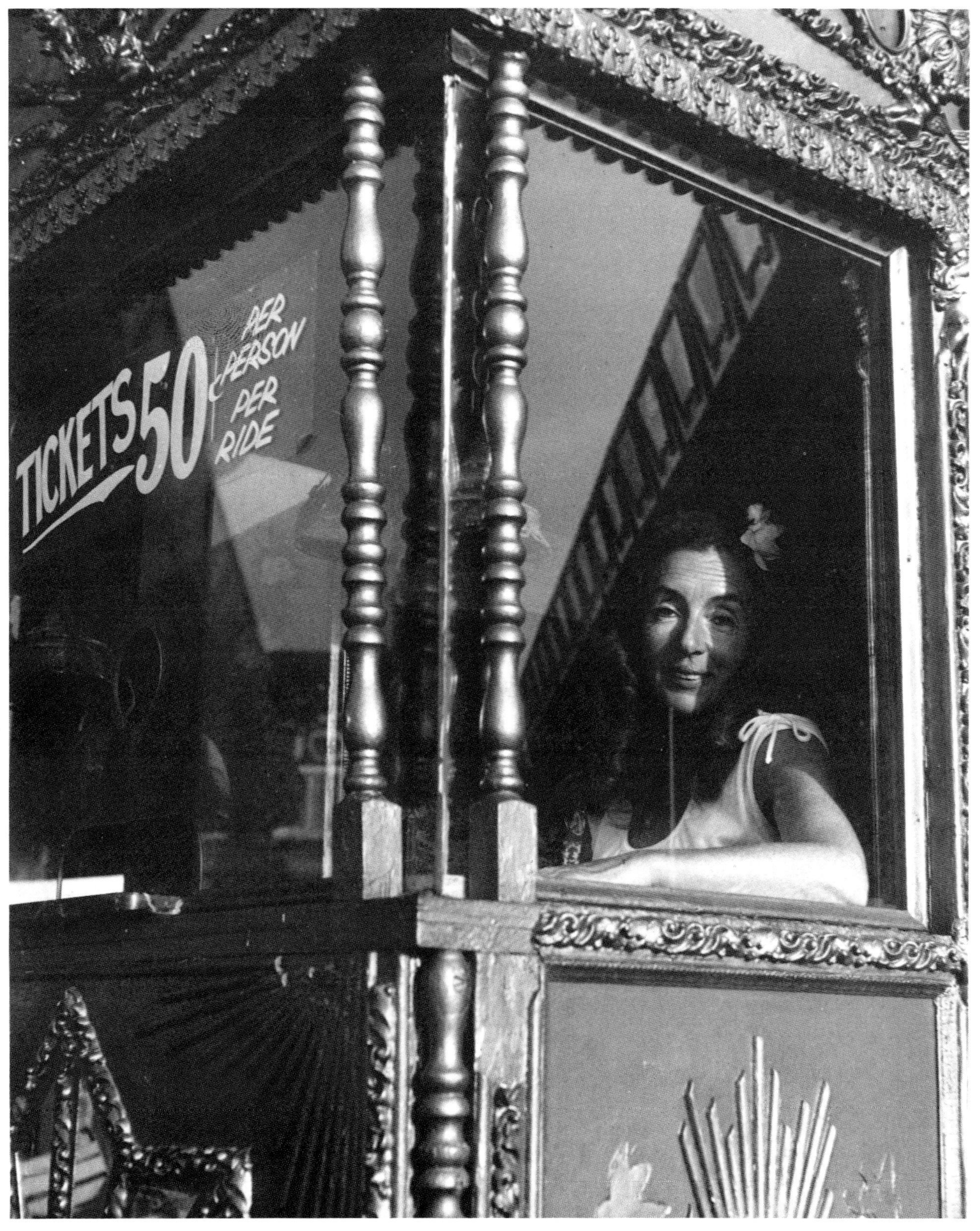

41. Yvonne

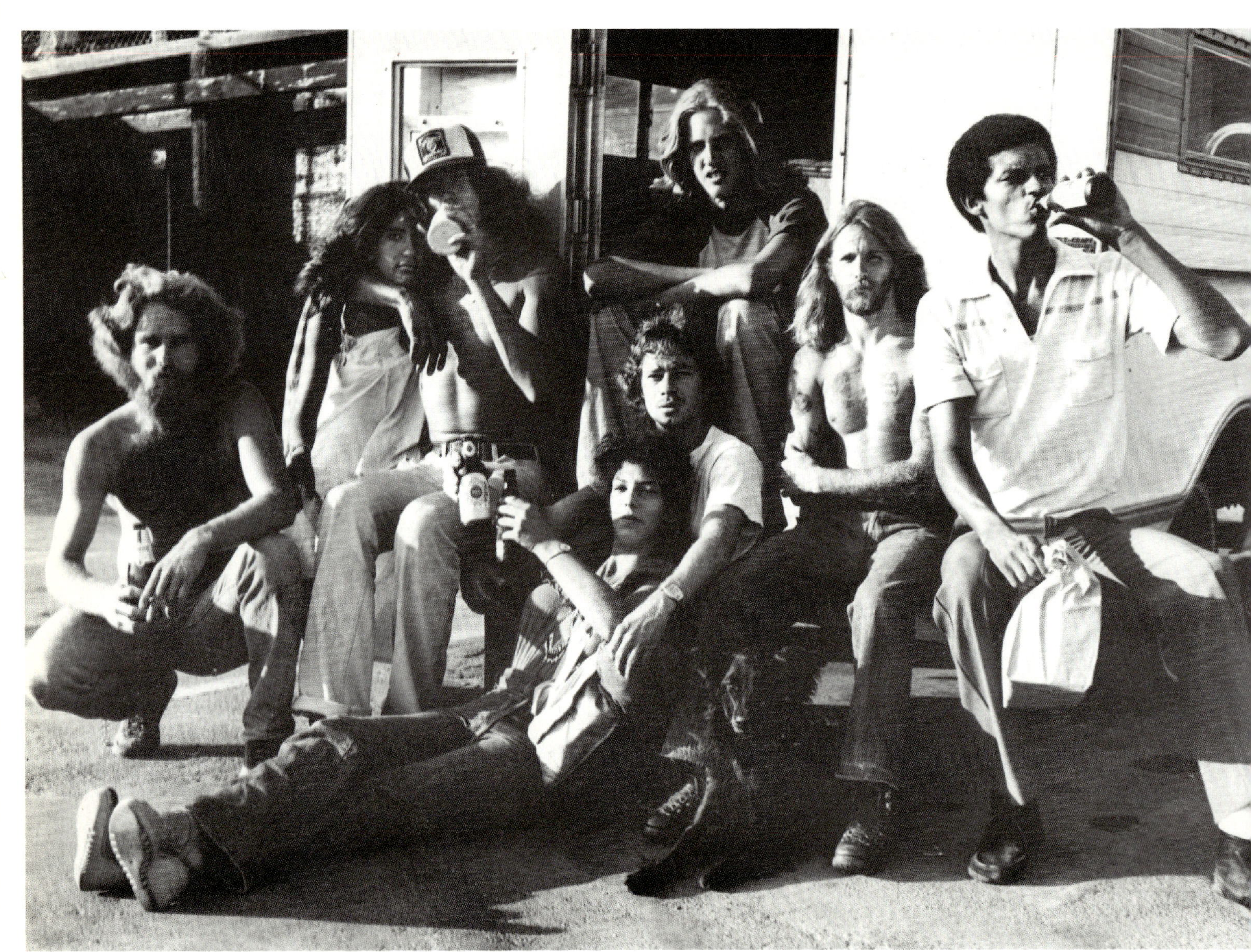

42. In the Parking Lot

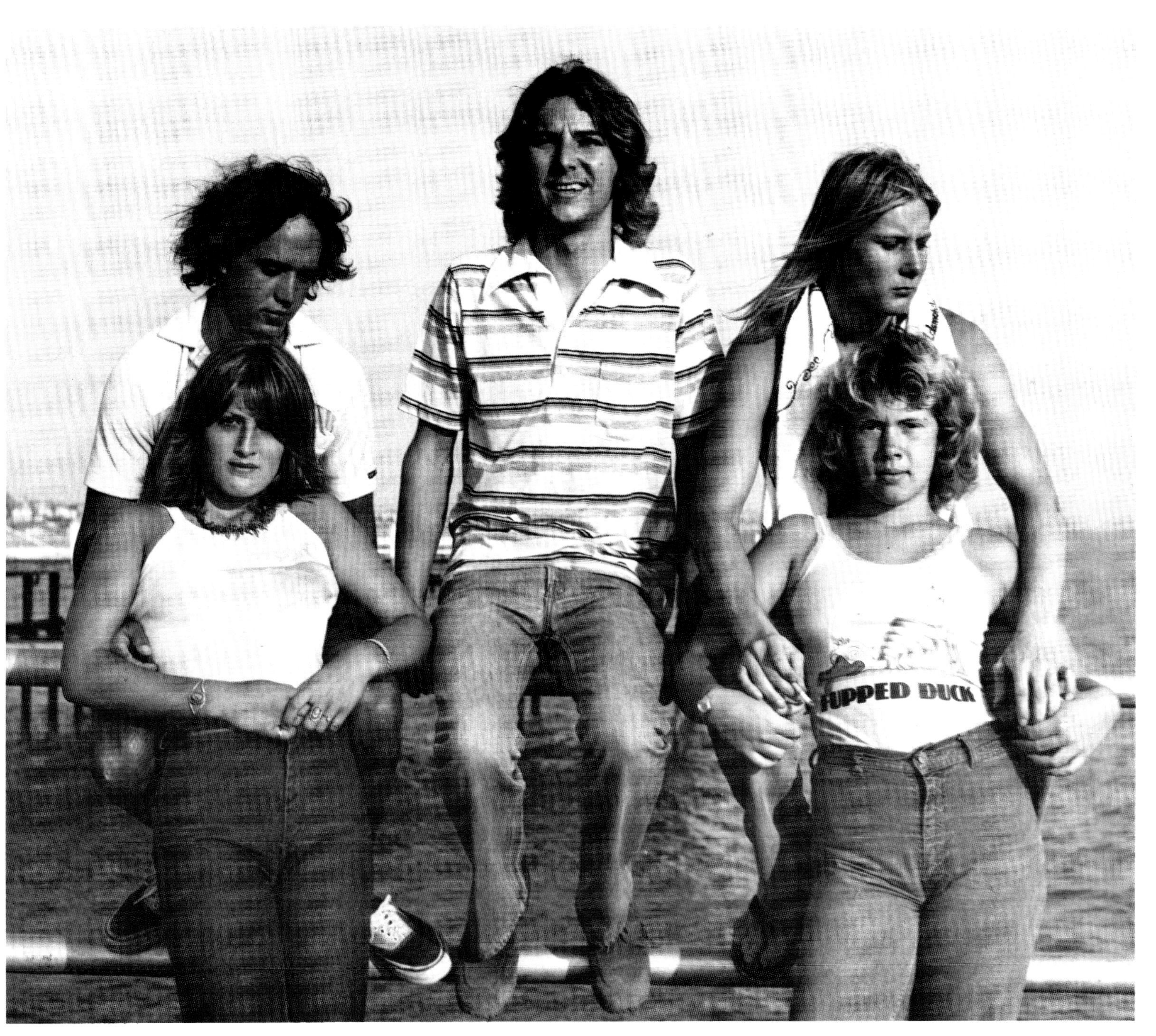

43. Jay and Cissy, Wayne, George and Dawn

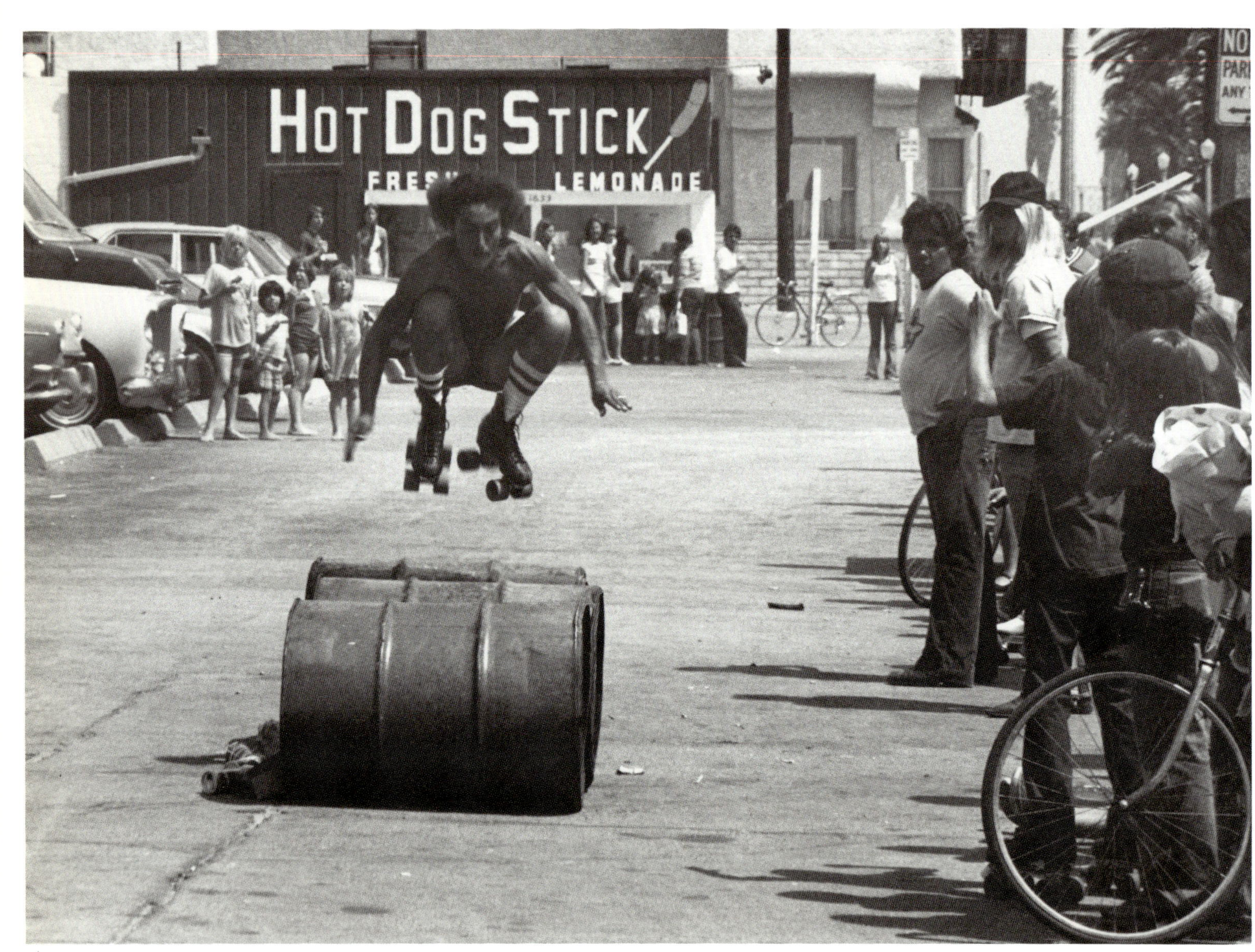

44. Ron Griego

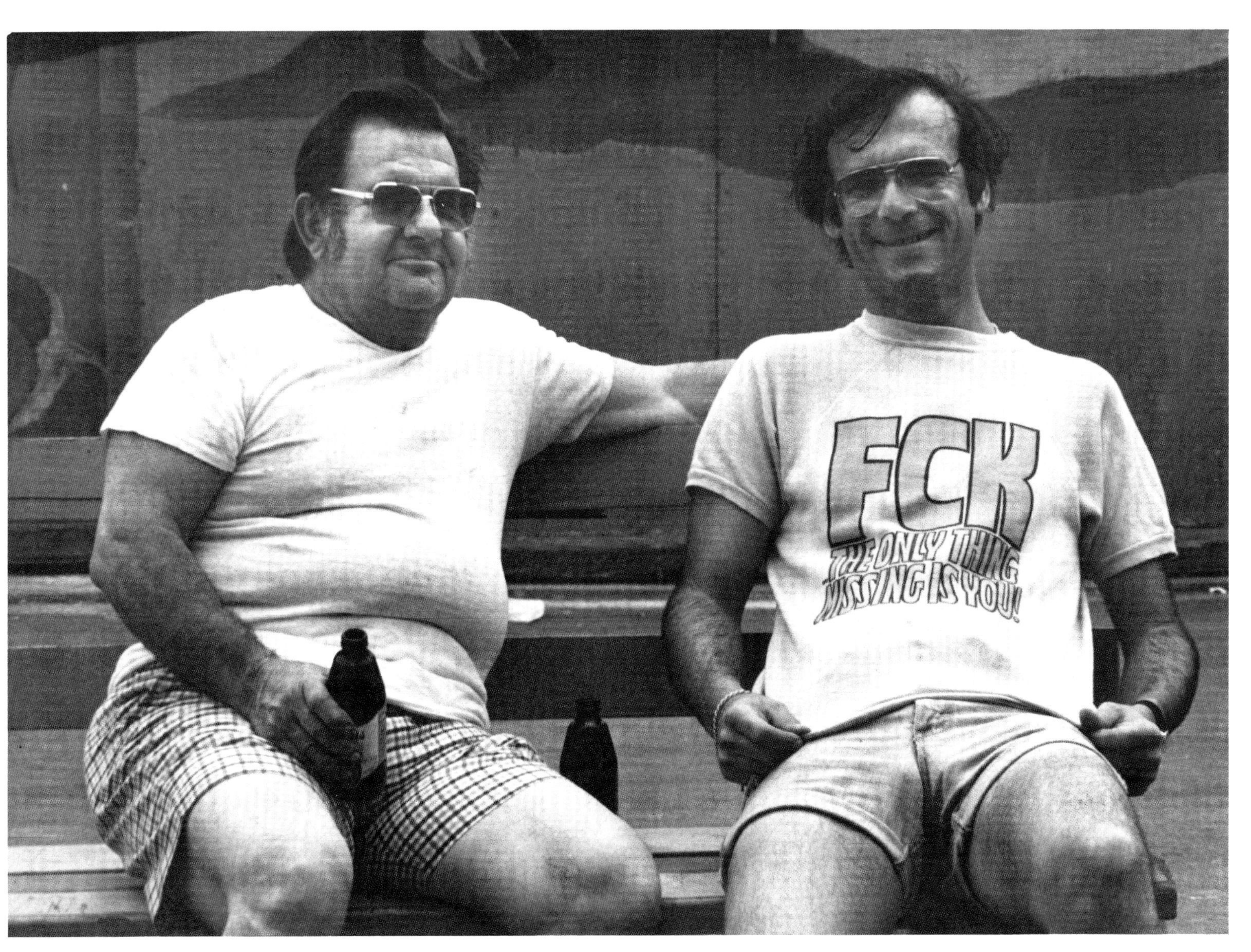

45. On the Boardwalk

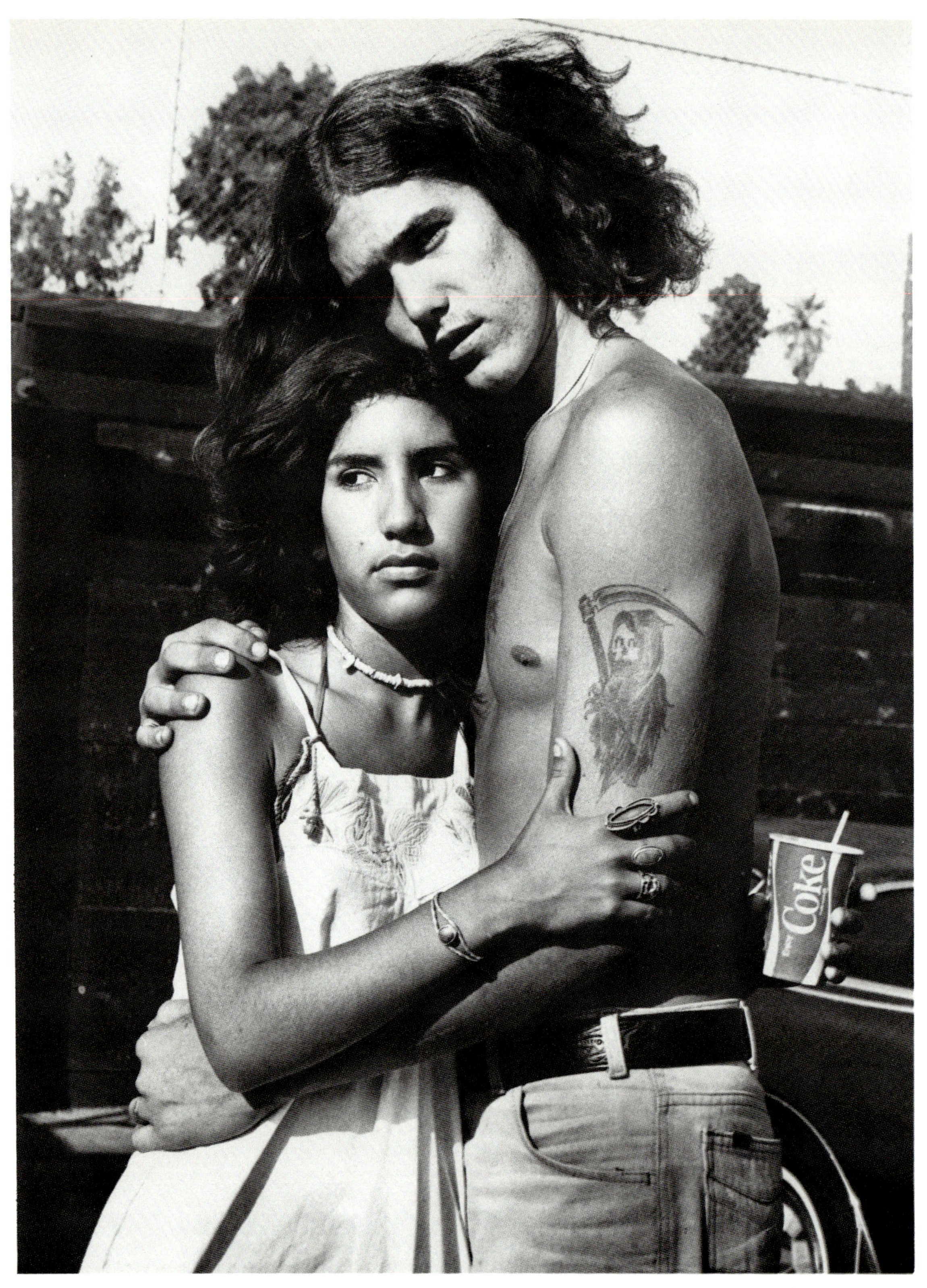

46. Slim Jim and Virginia

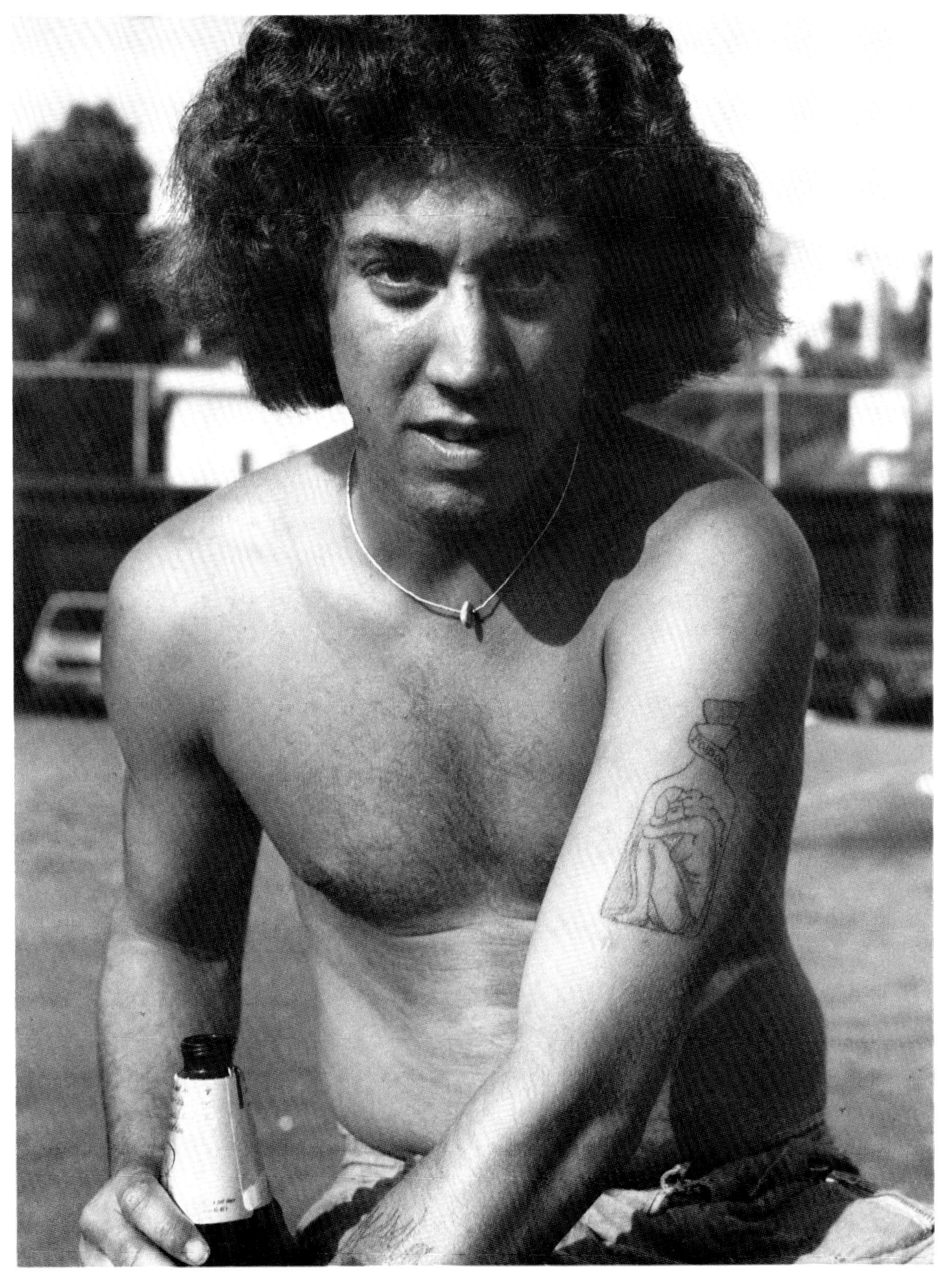

47. Crispy Cory

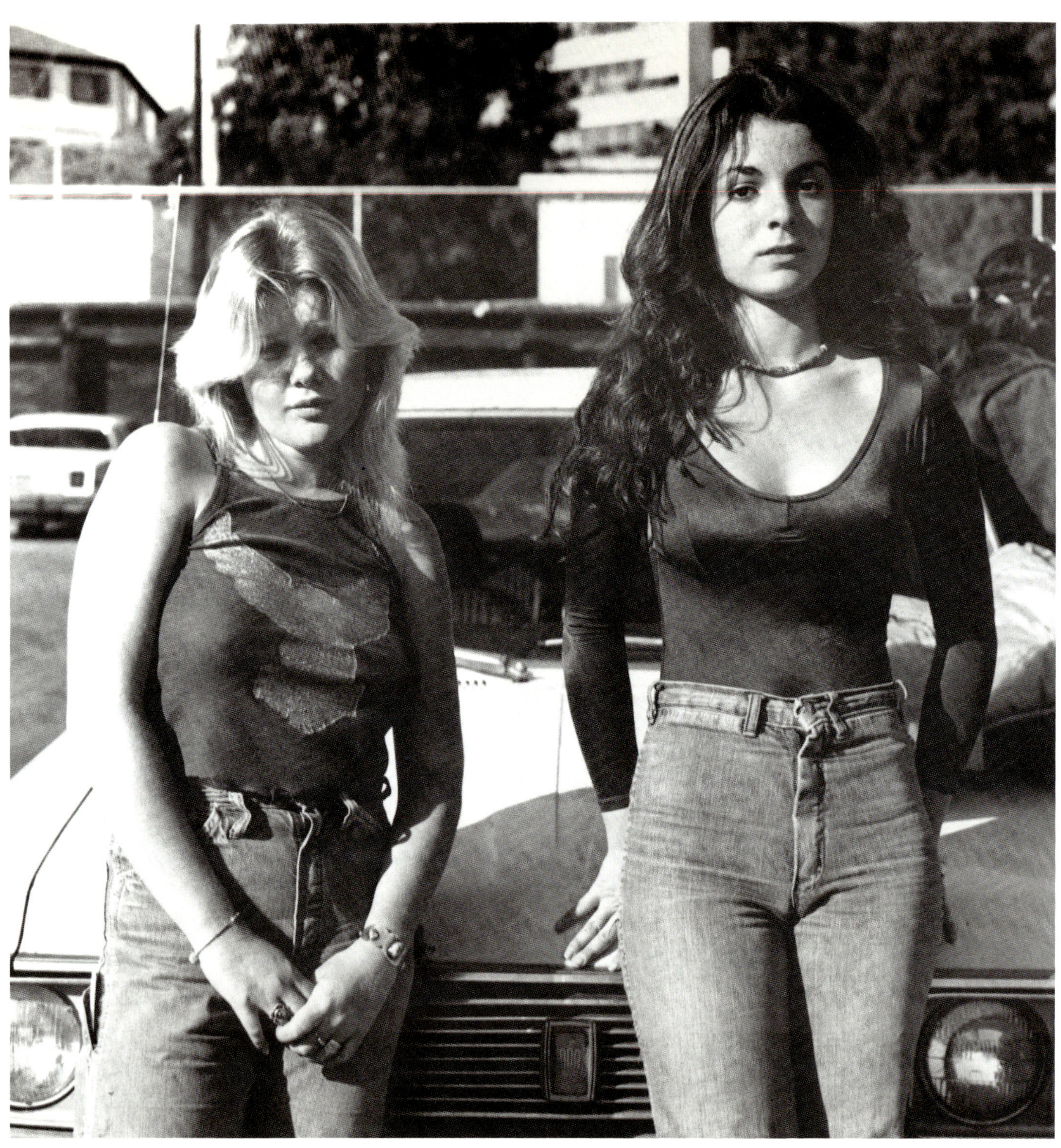

48. Mary and Beauty

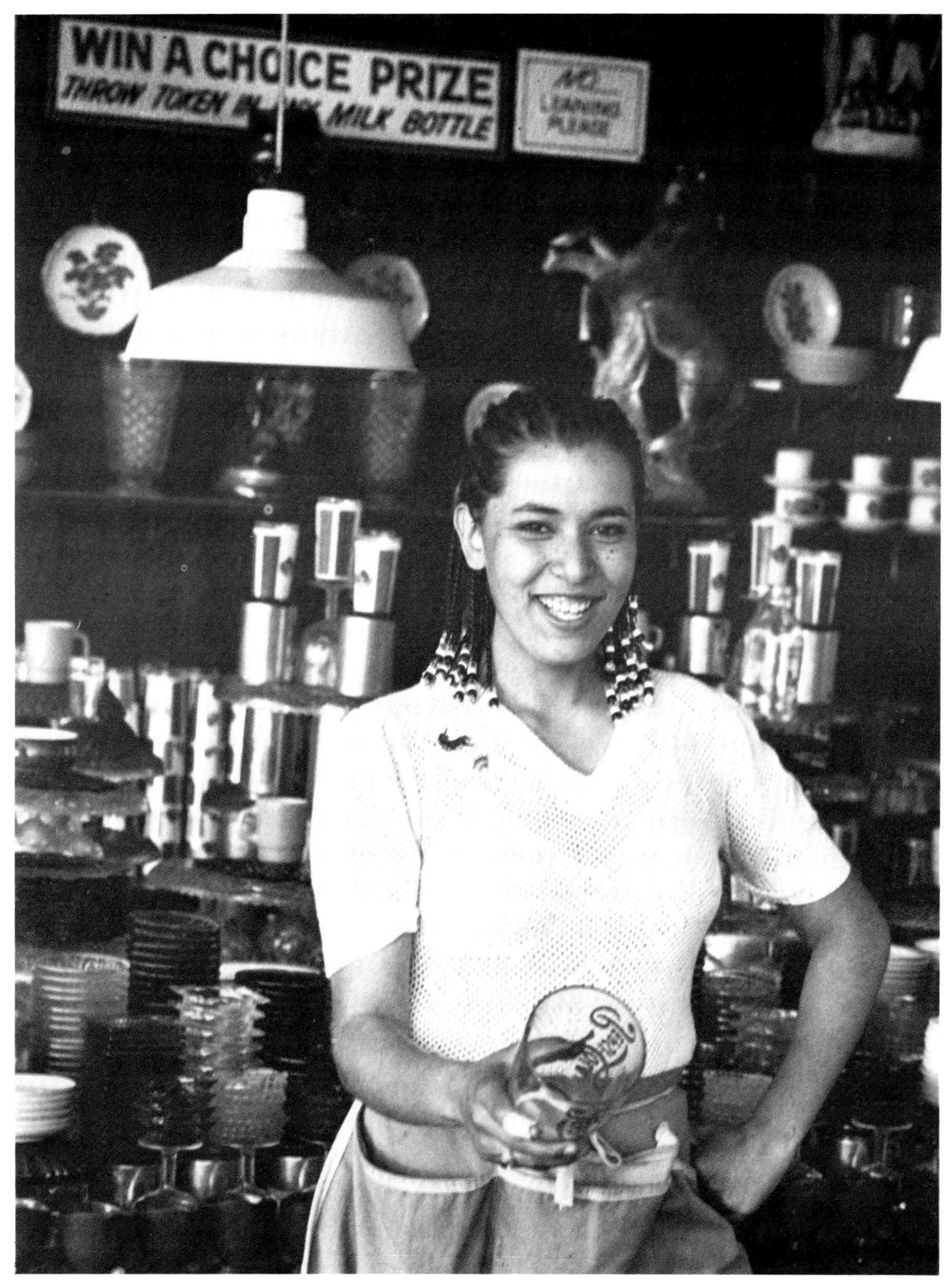

49. Laura at the Dime Toss

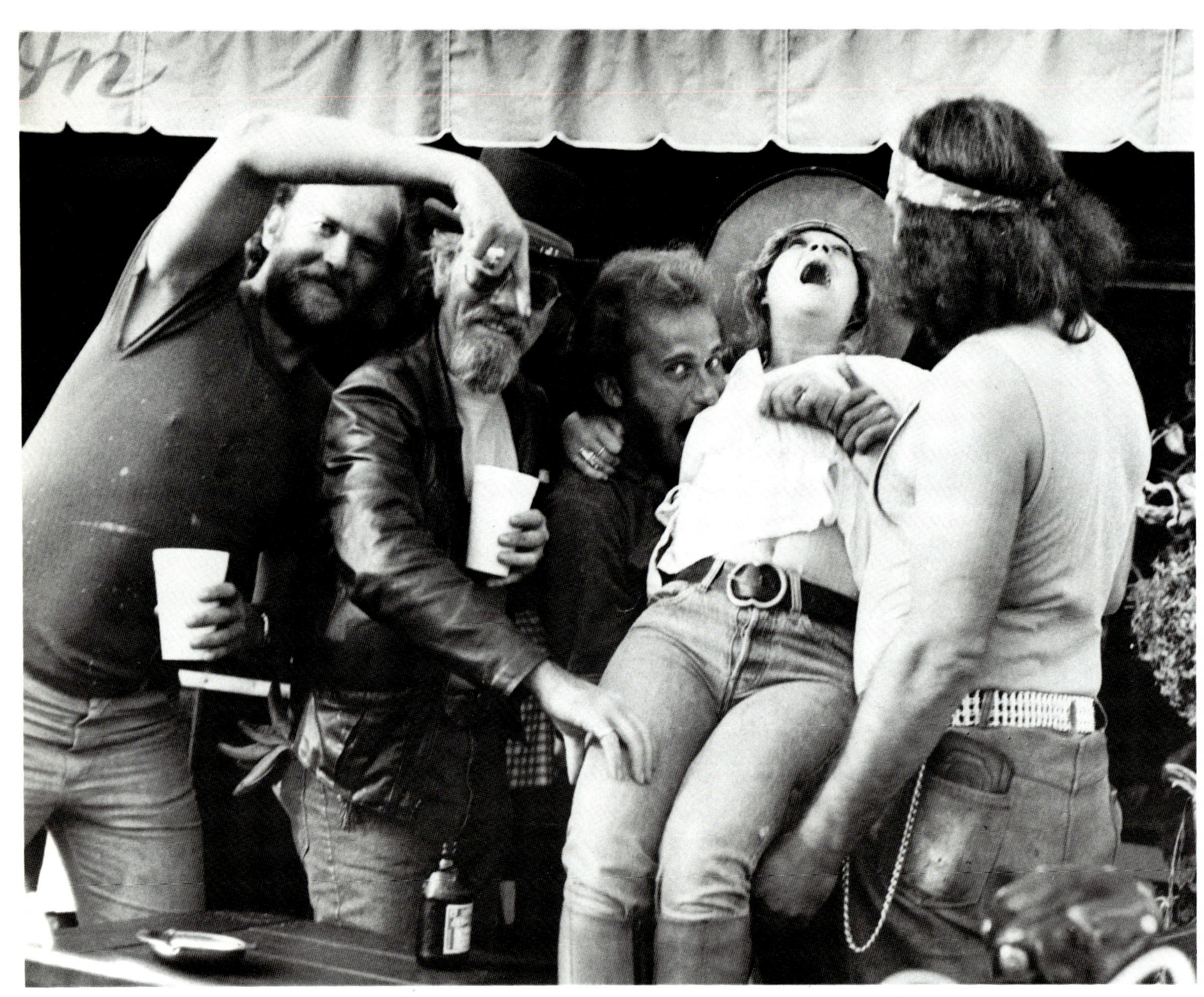

50. Sean, Jerry, Randy, Sandy, and Mac

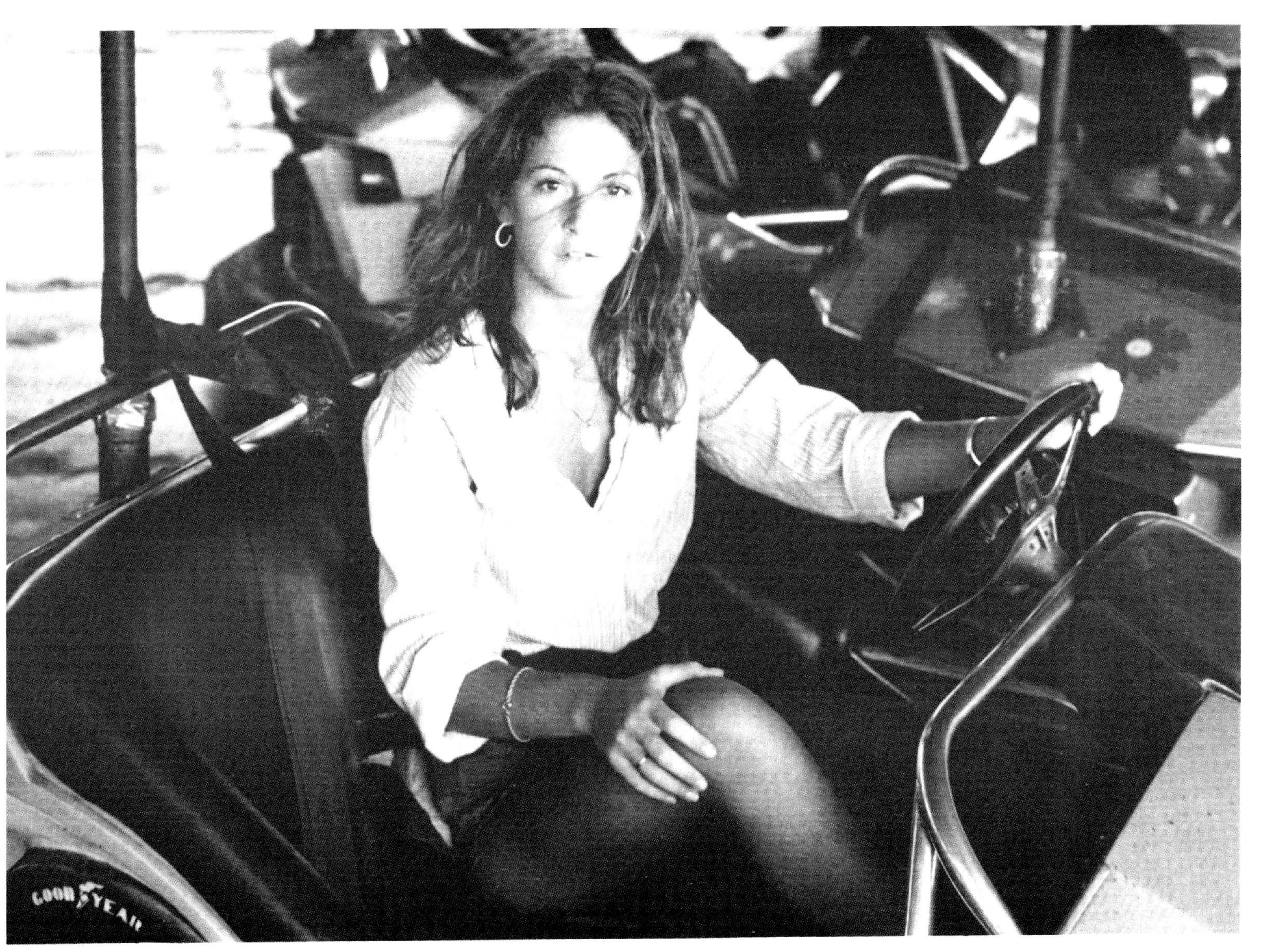

51. Lolita

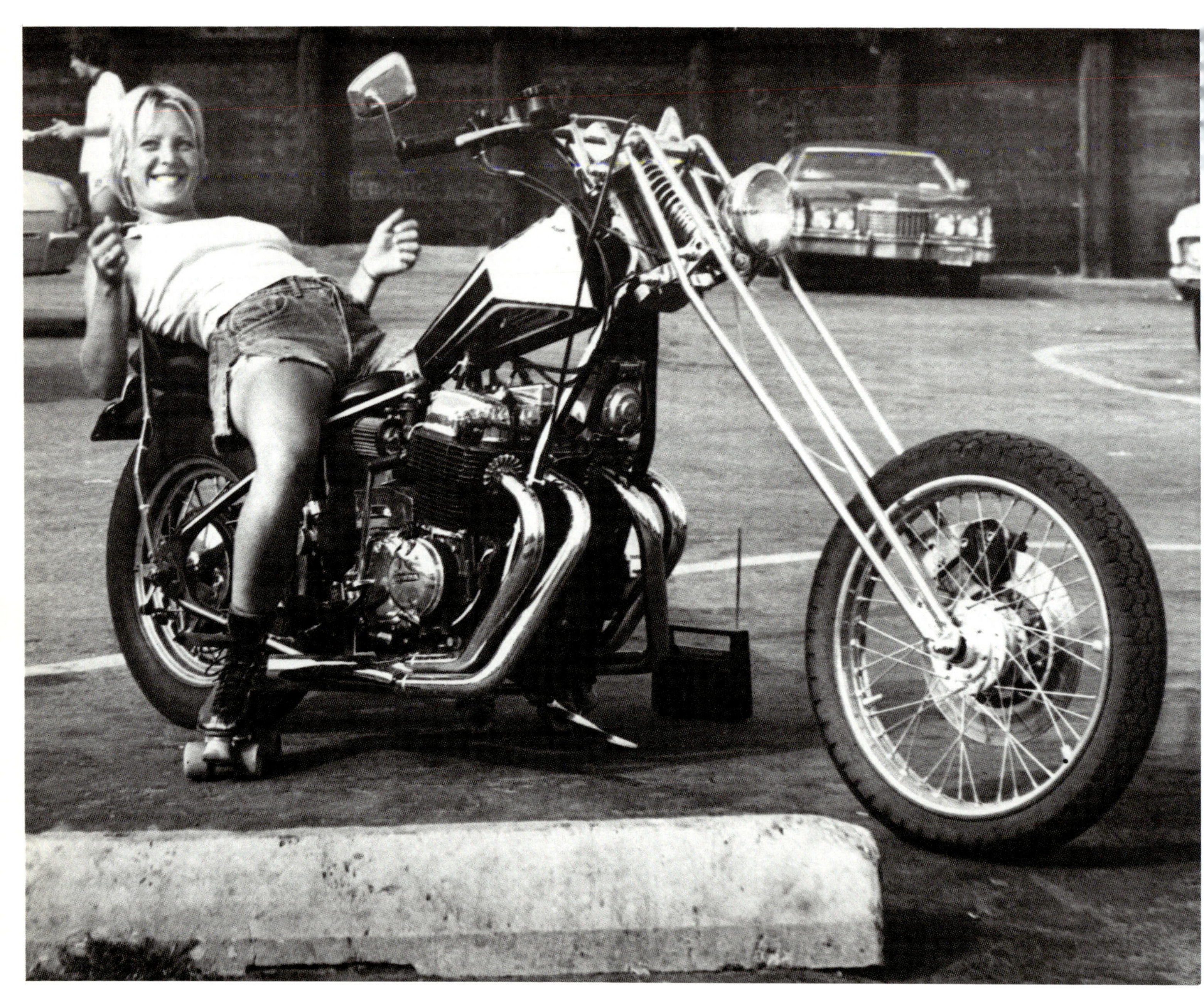

52. On Jerry's Bike

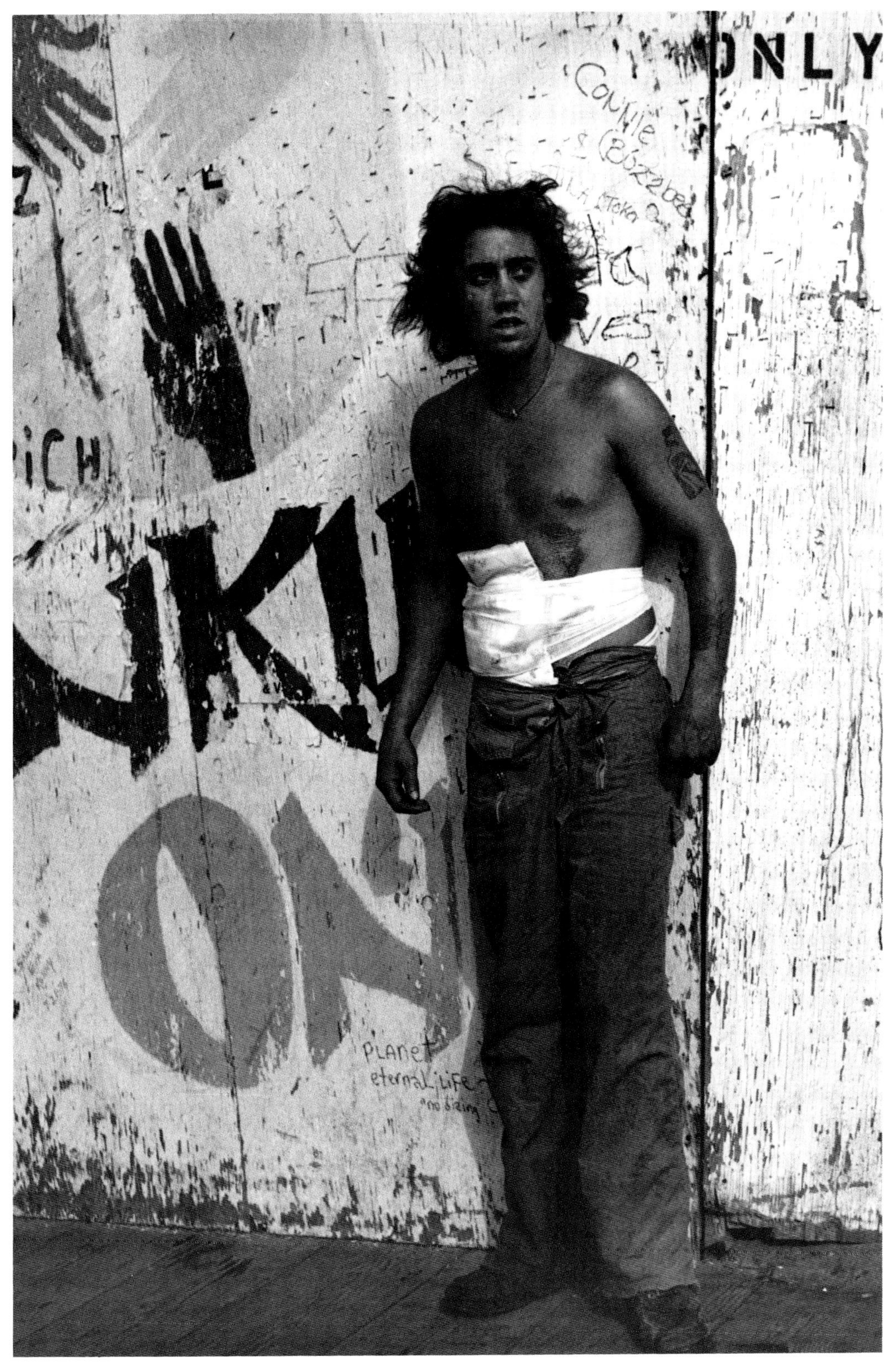

53. Crispy Cory, After the Accident

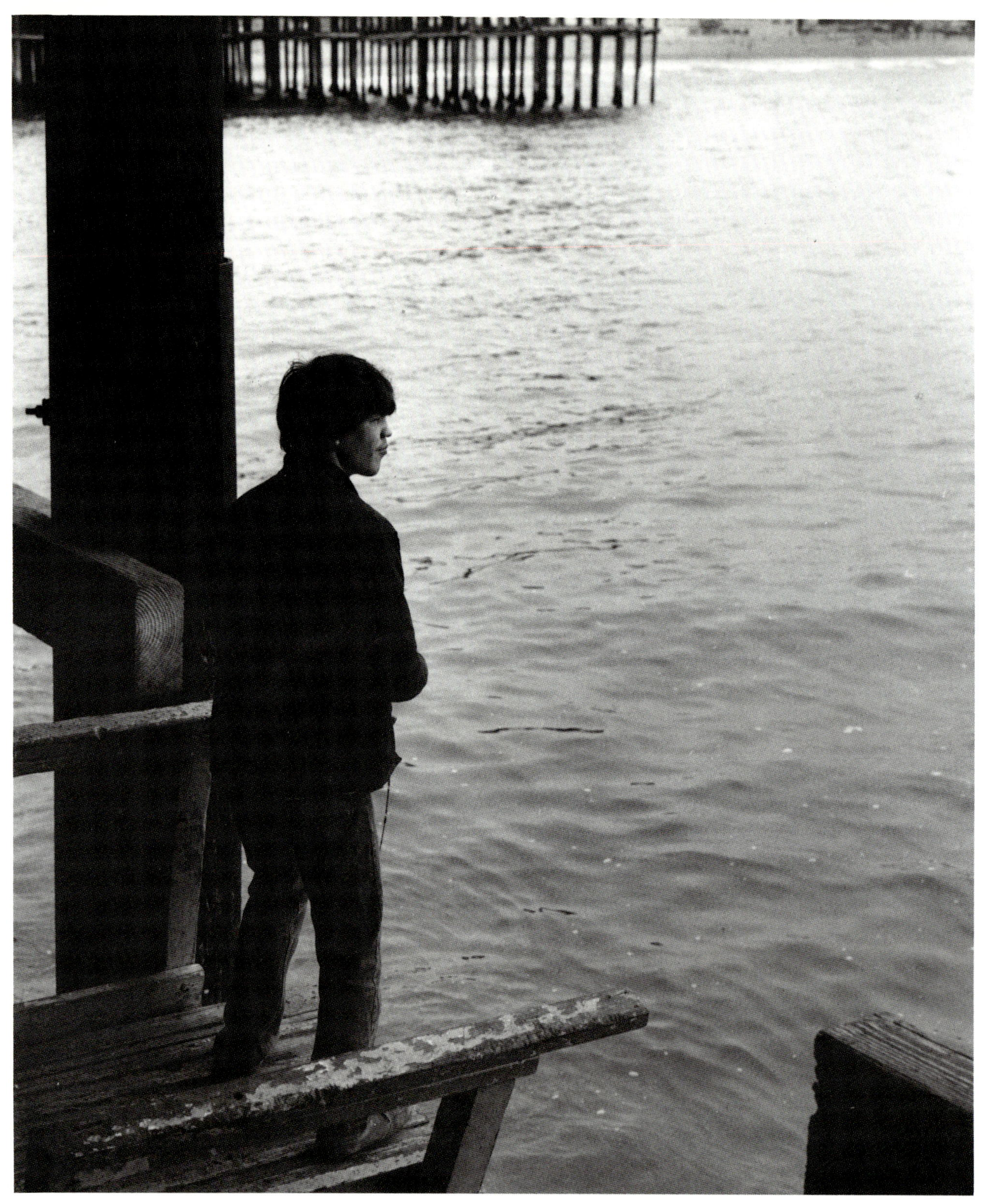

54. Lower Pier

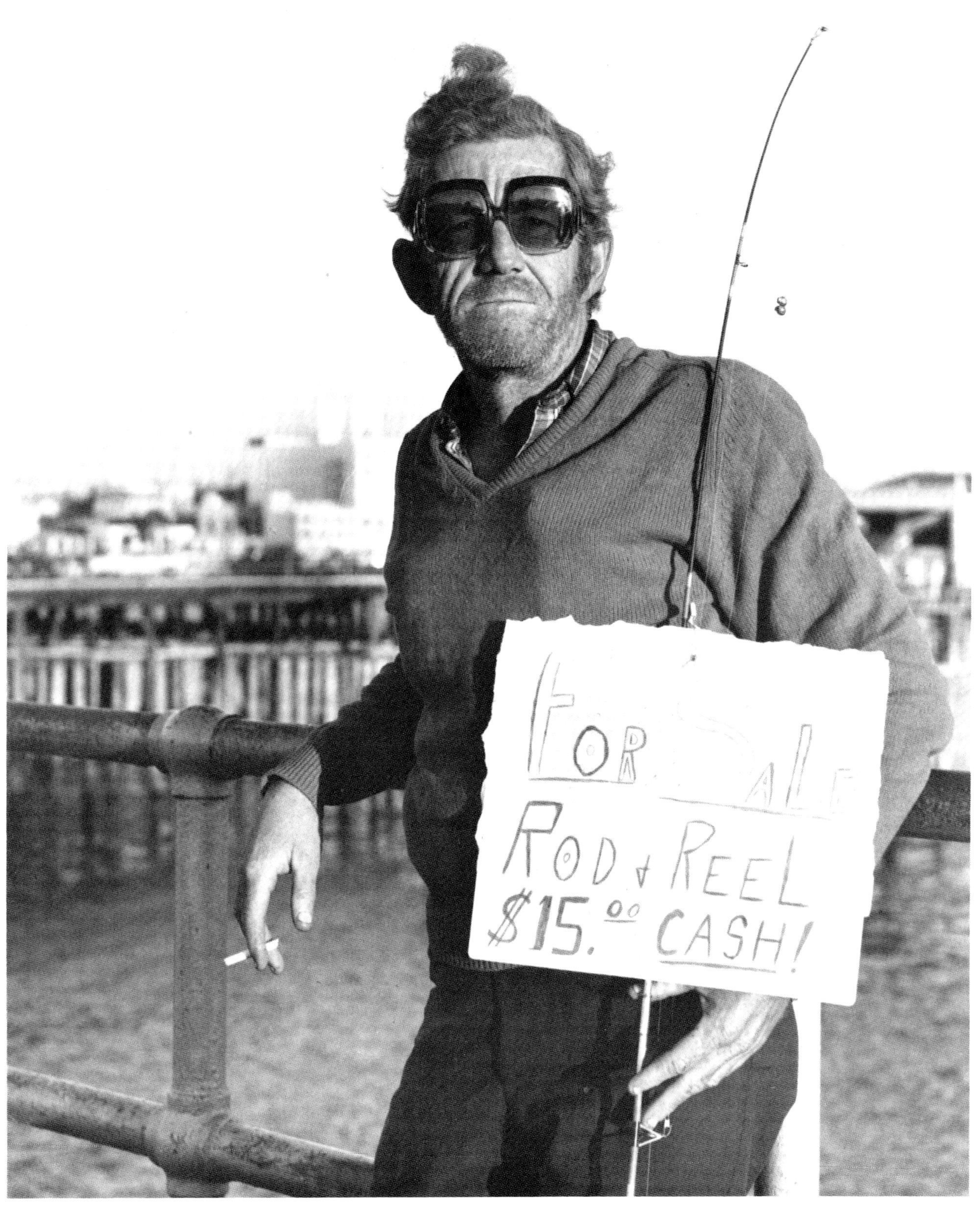

55. *Don*

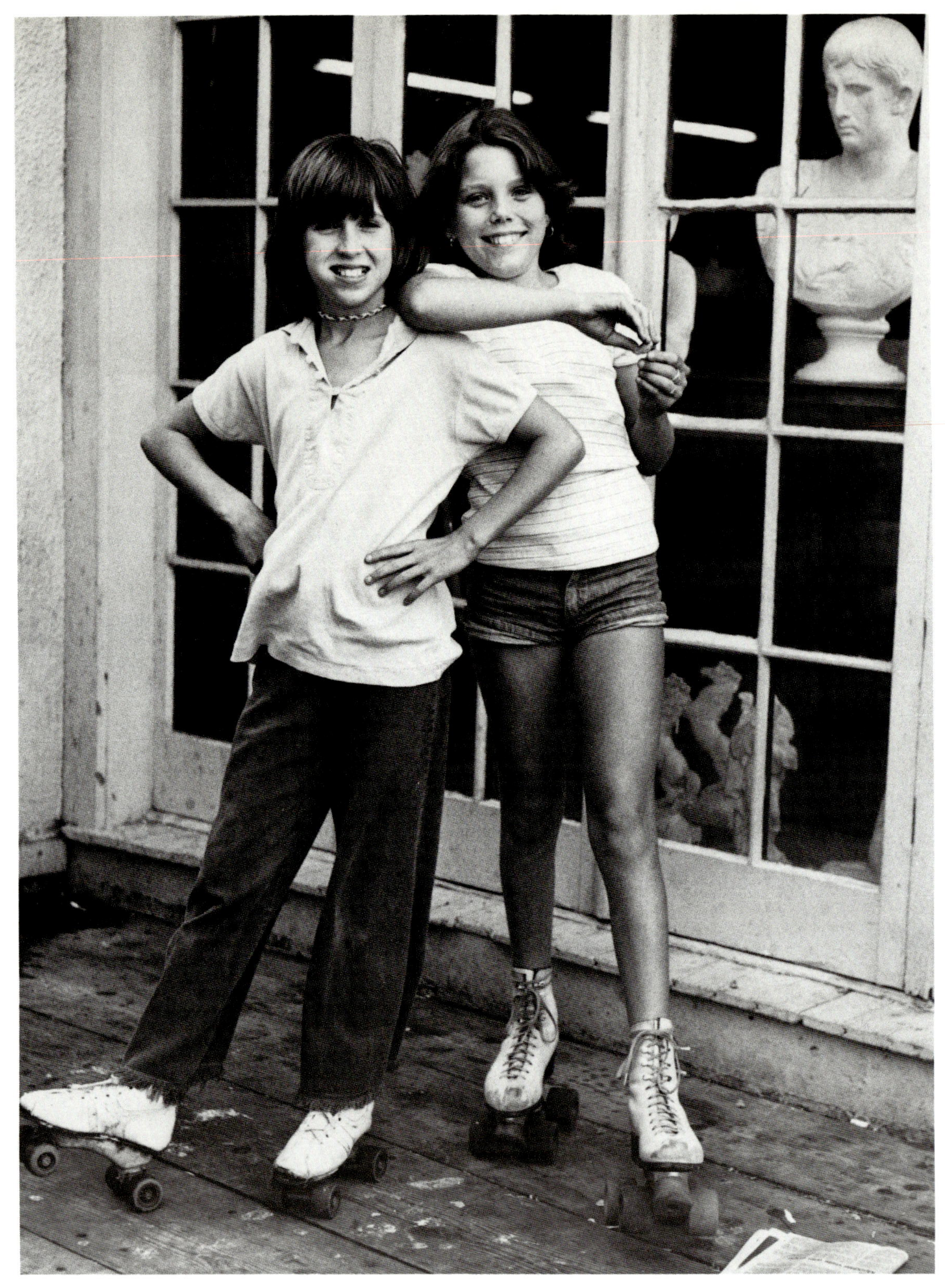

56. Angel and Aaron

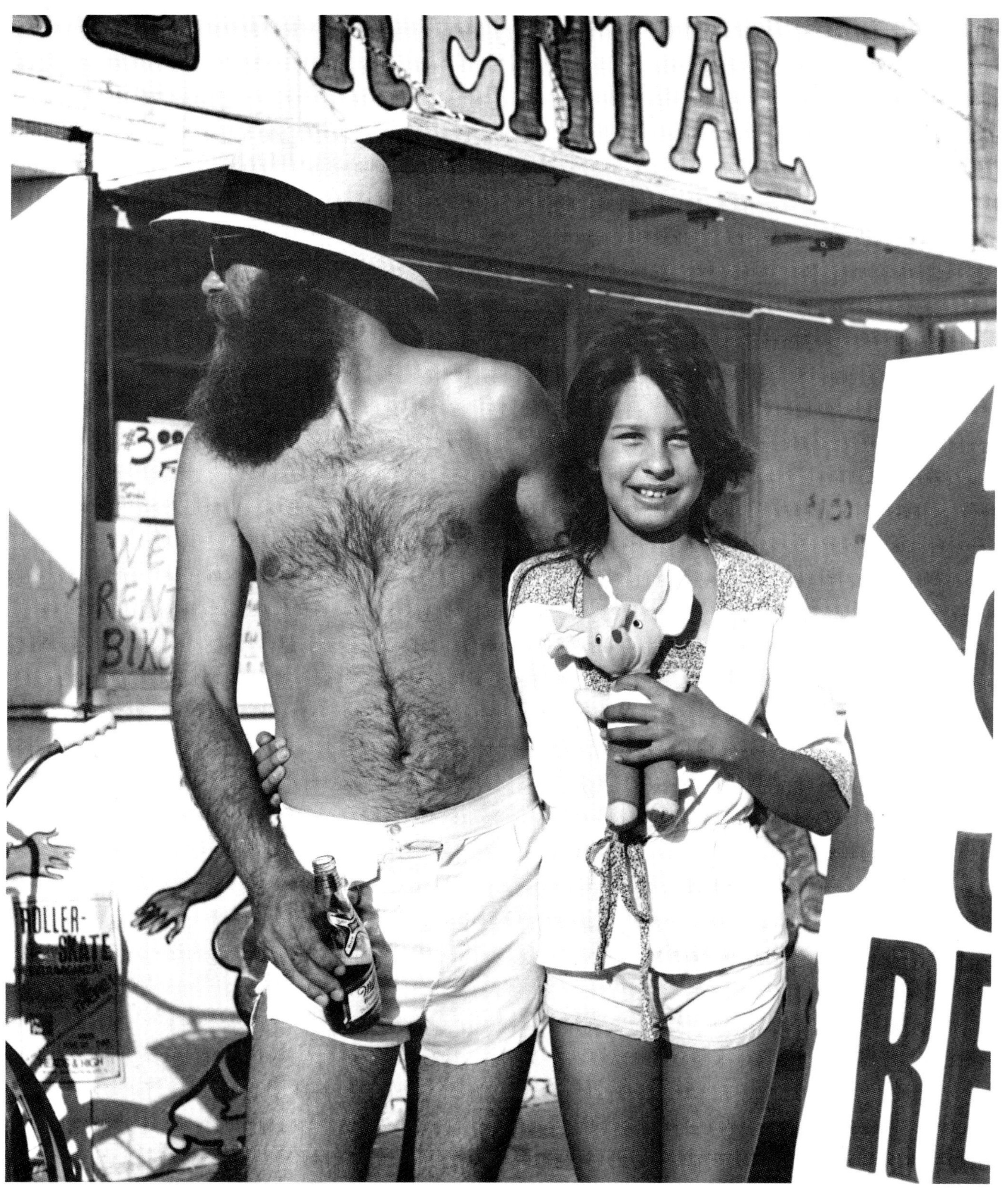

57. Charlie and Lisa

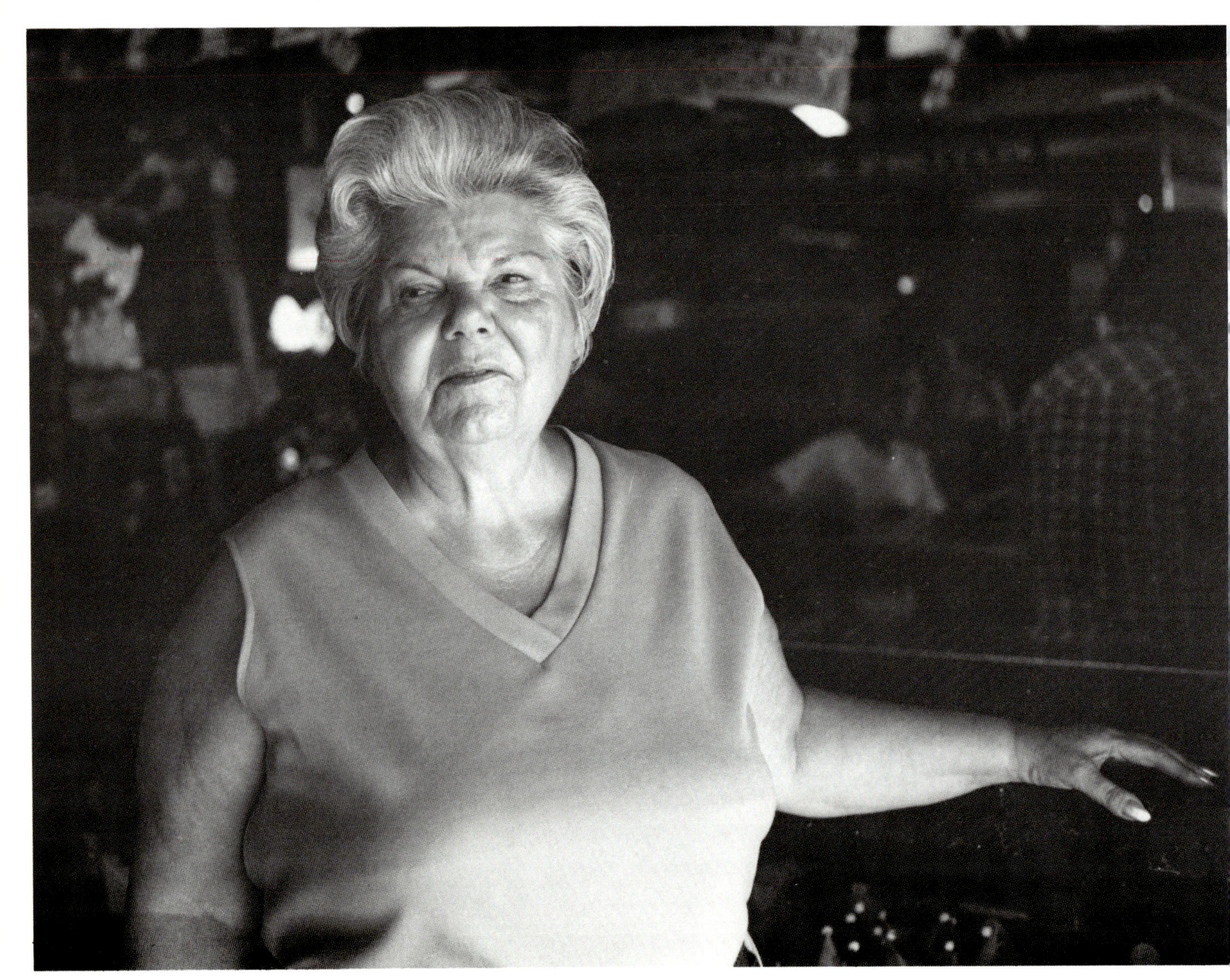

58. May McCabe

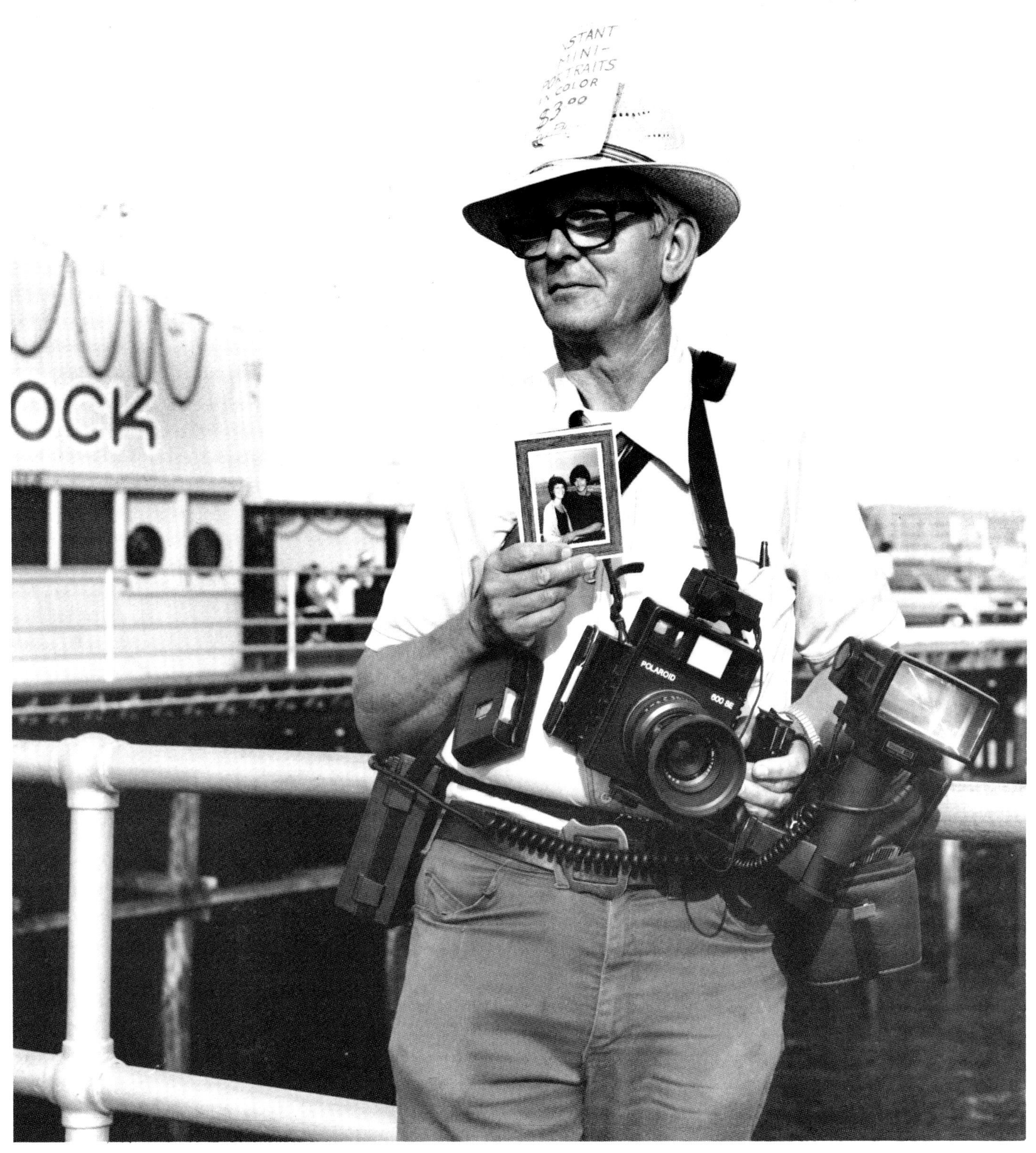

59. Al the Photographer

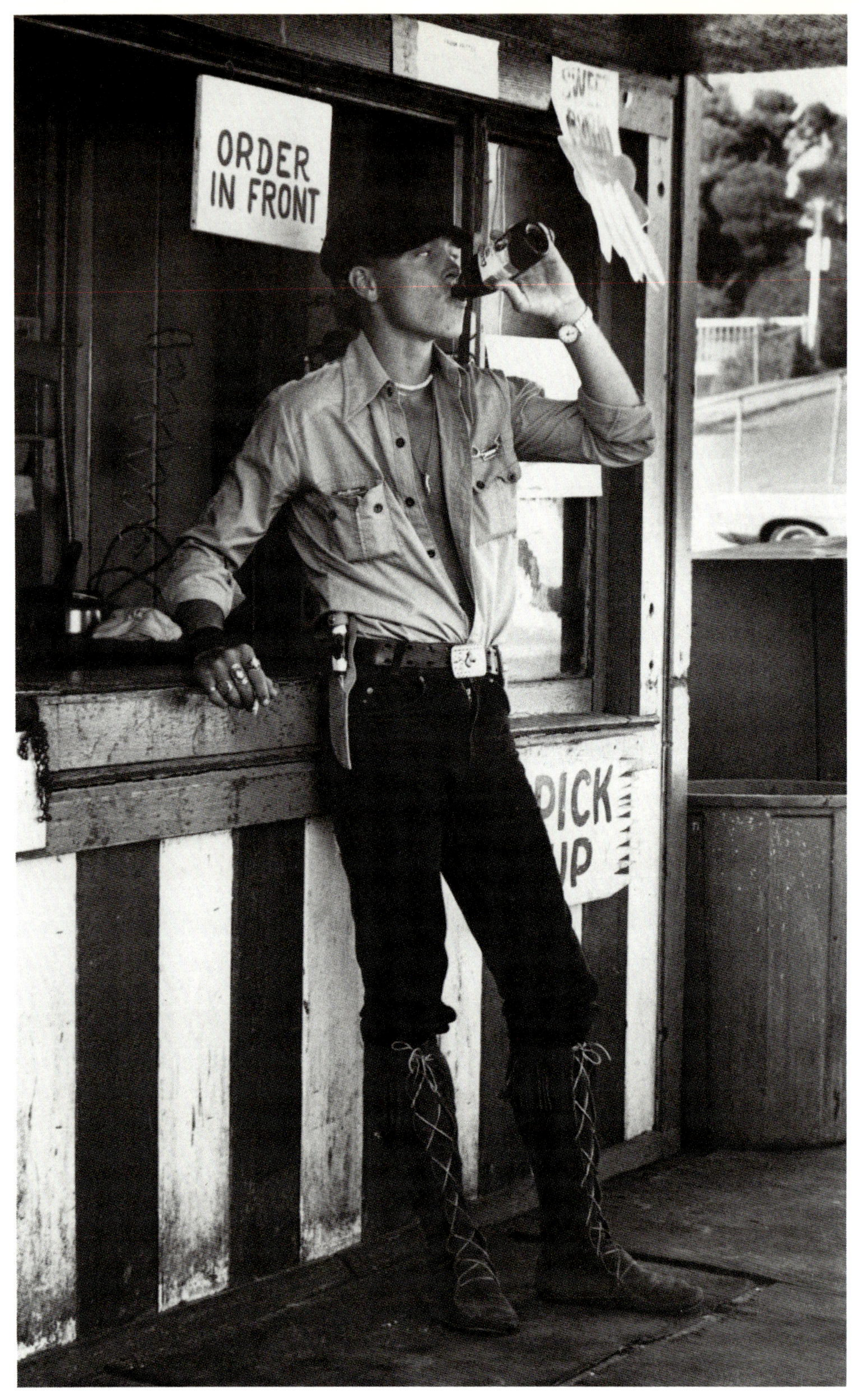

60. Doug

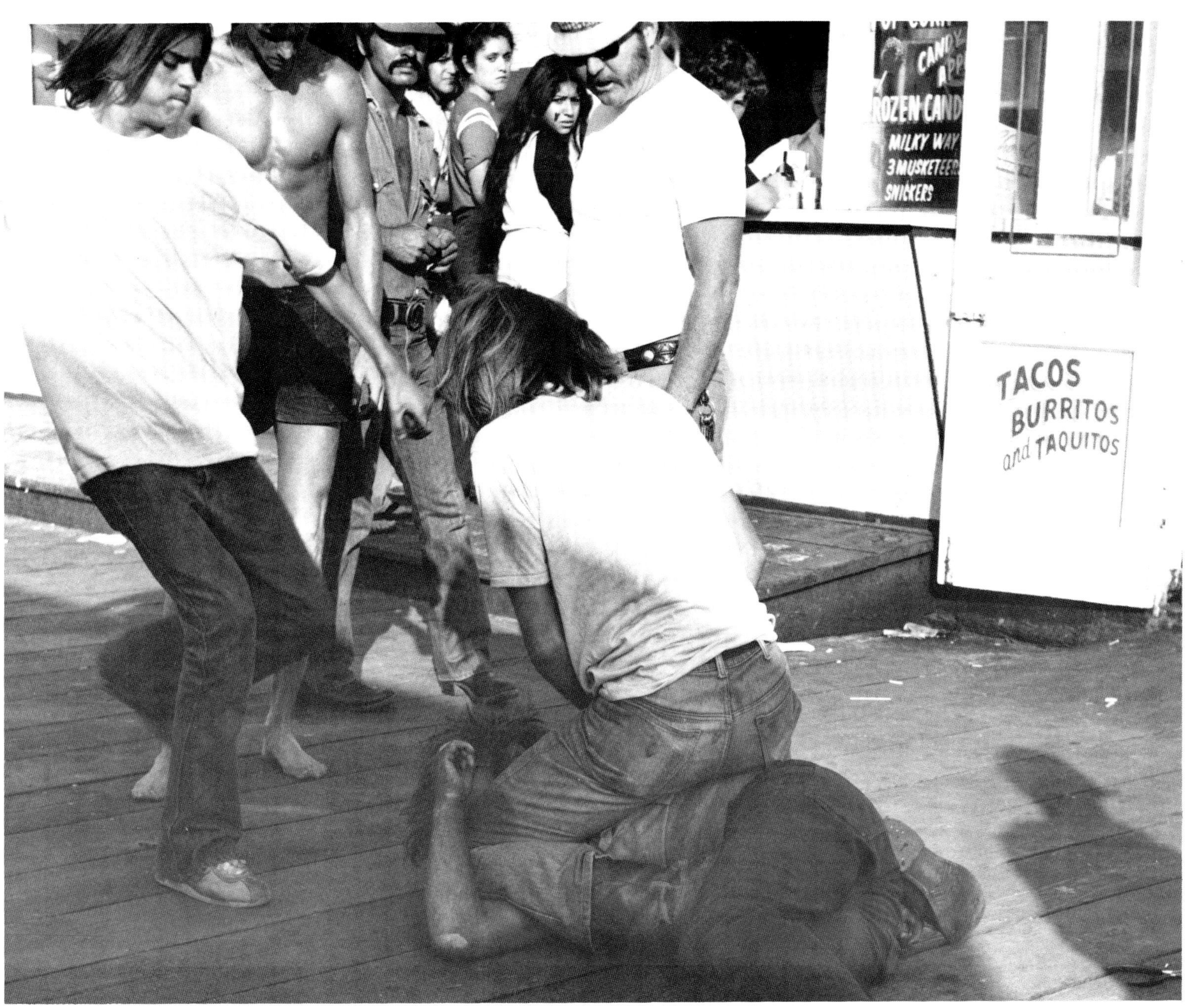

61. The Beating

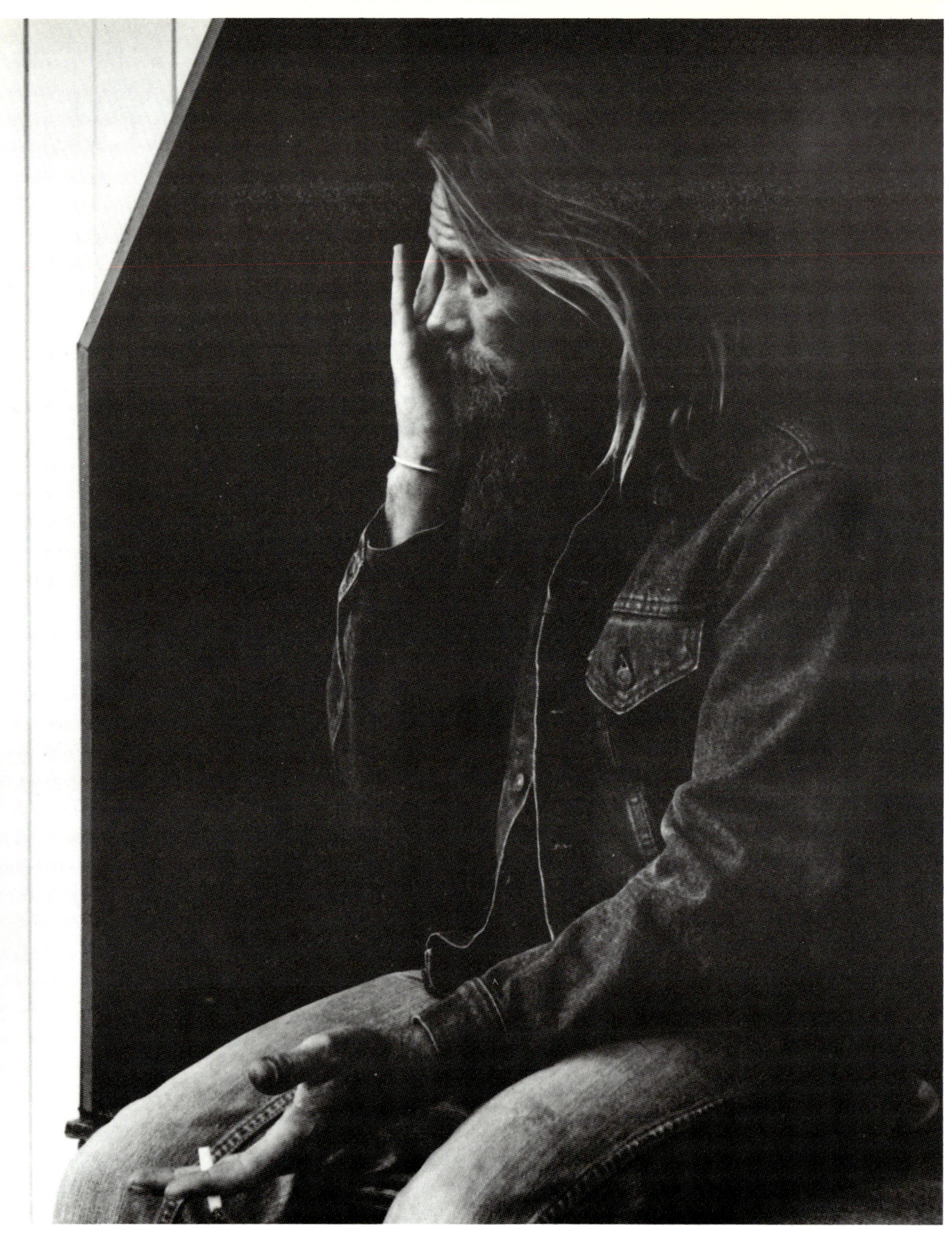

62. Wild Bill

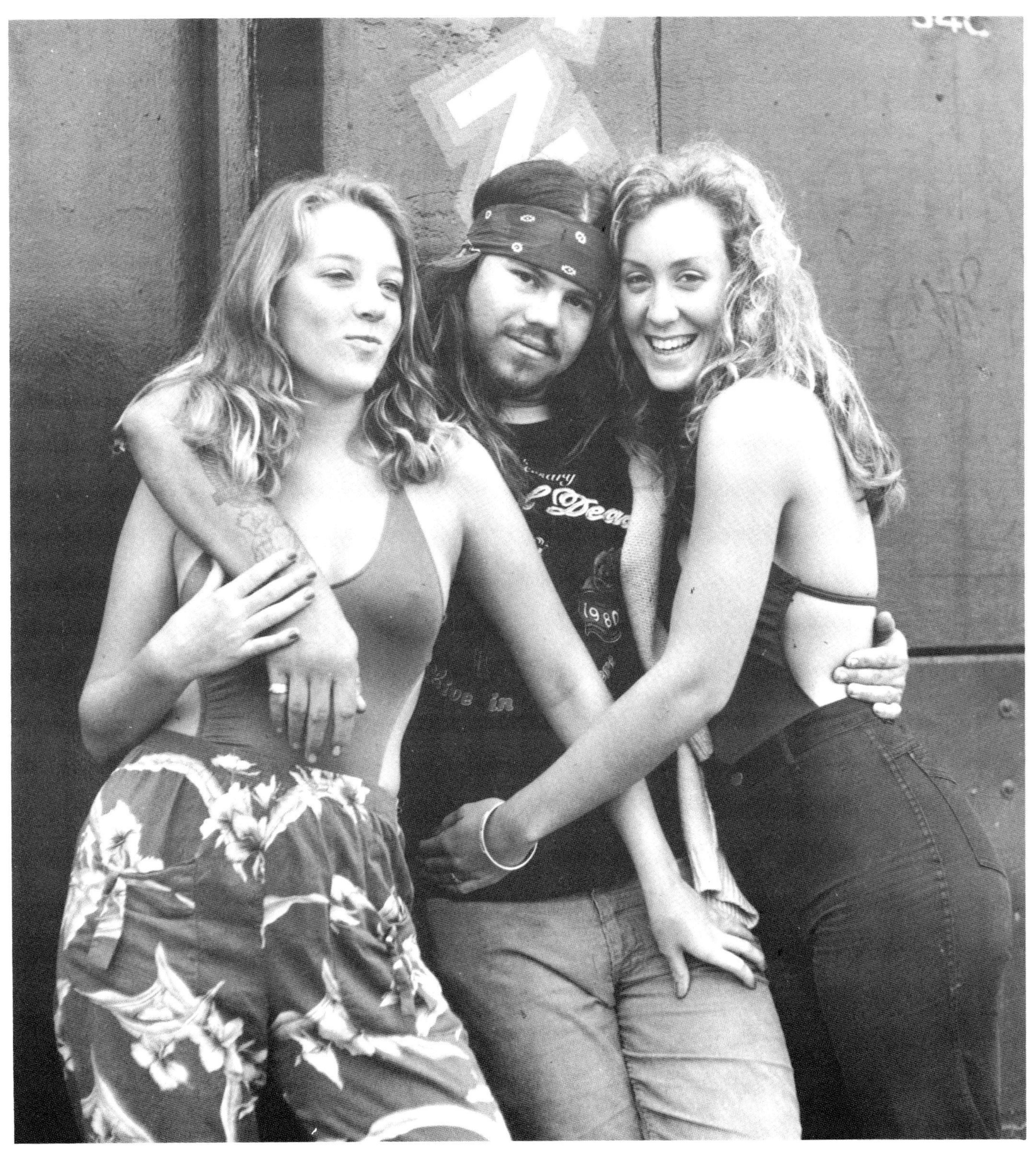

63. Nancy, Critter, and Matina

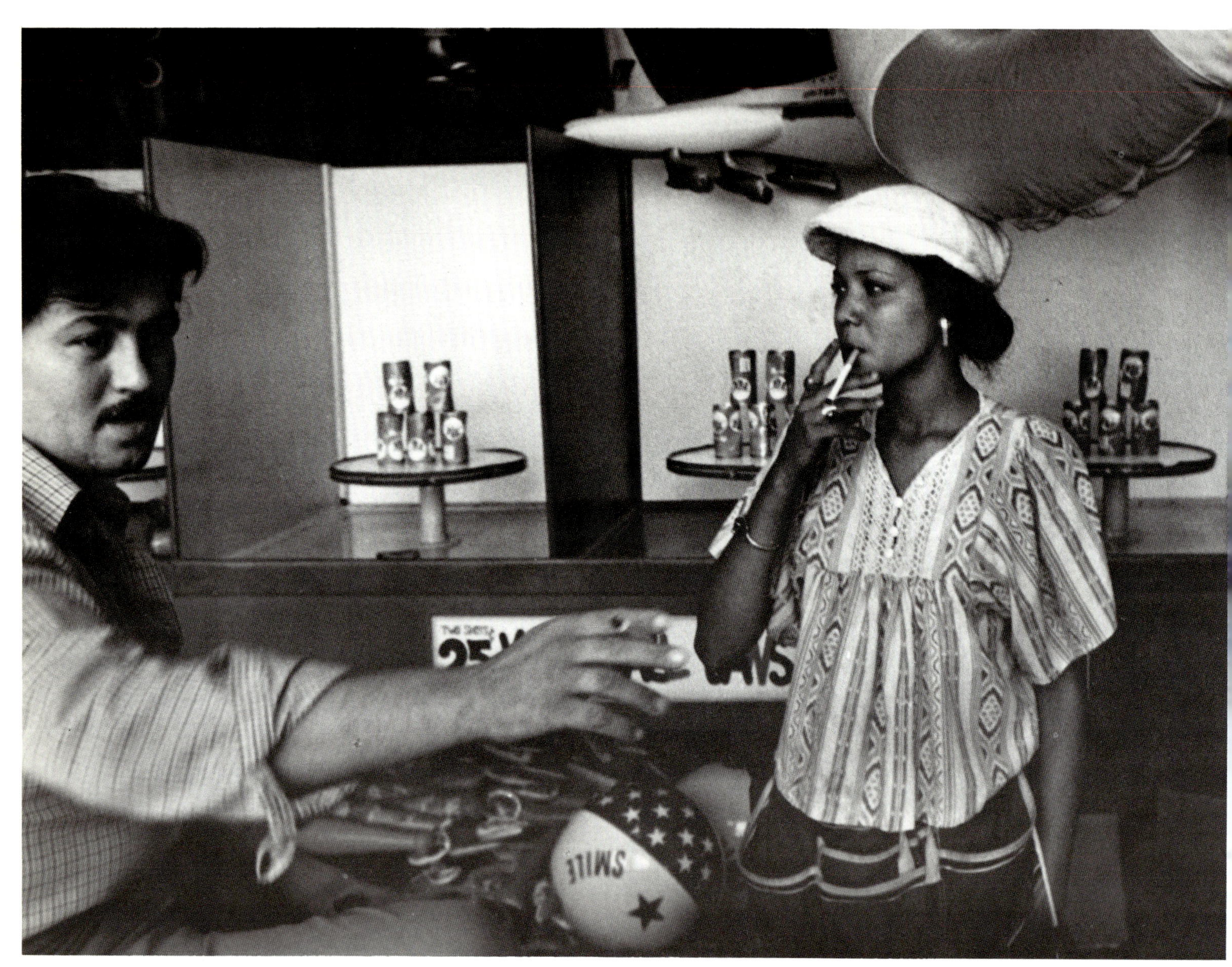

64. Tasha and the Manager

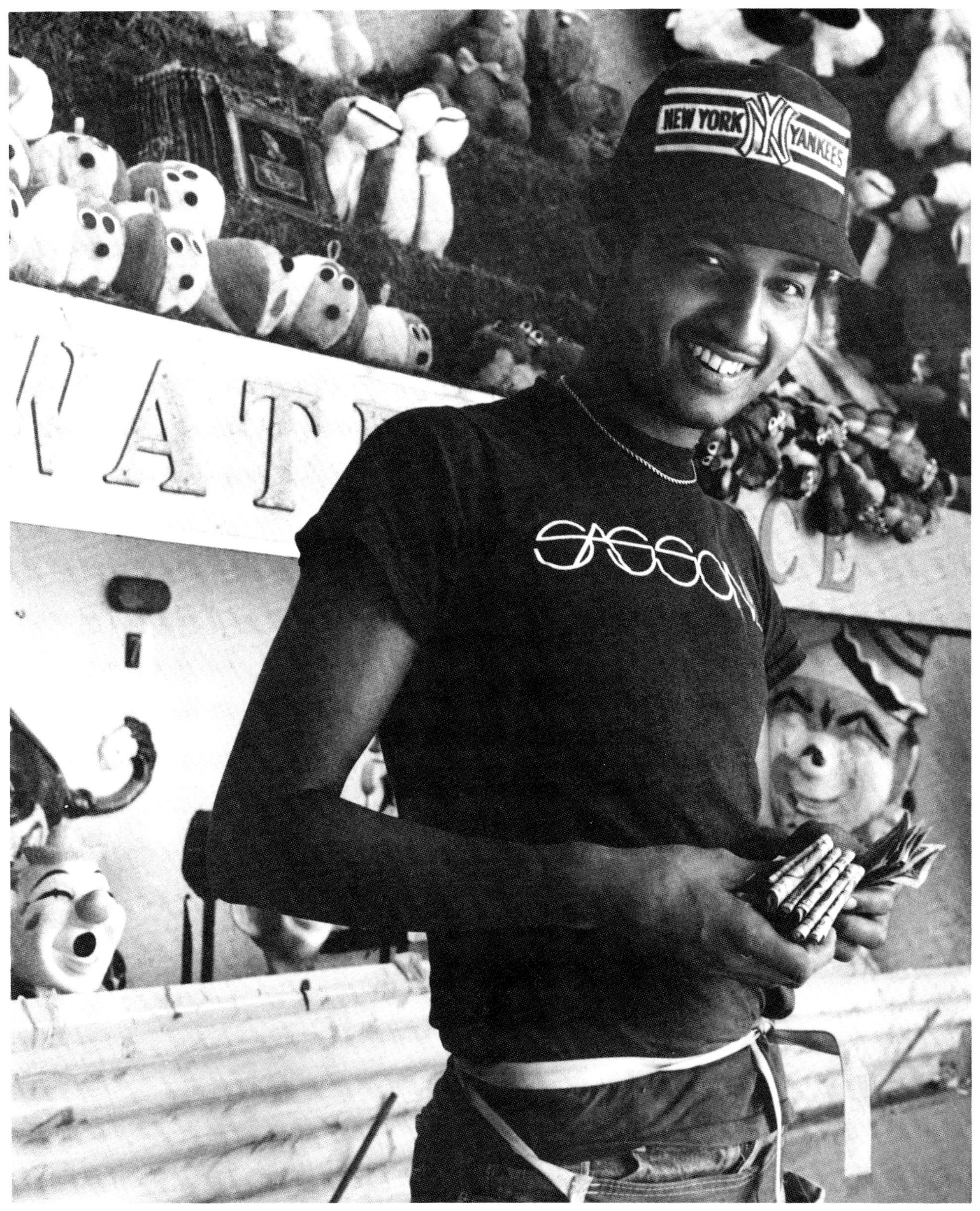

65. Mario

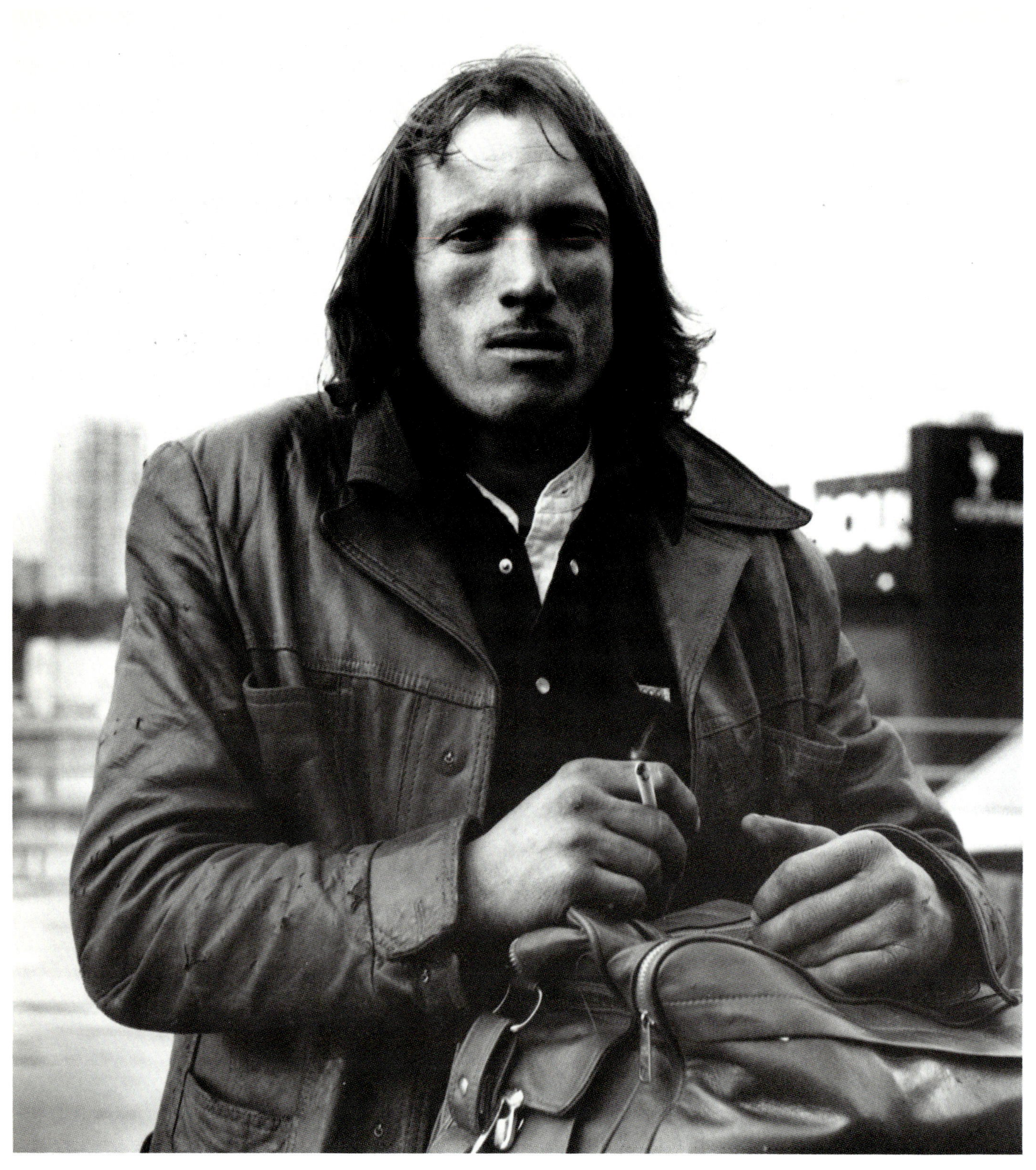

66. Man in a Storm

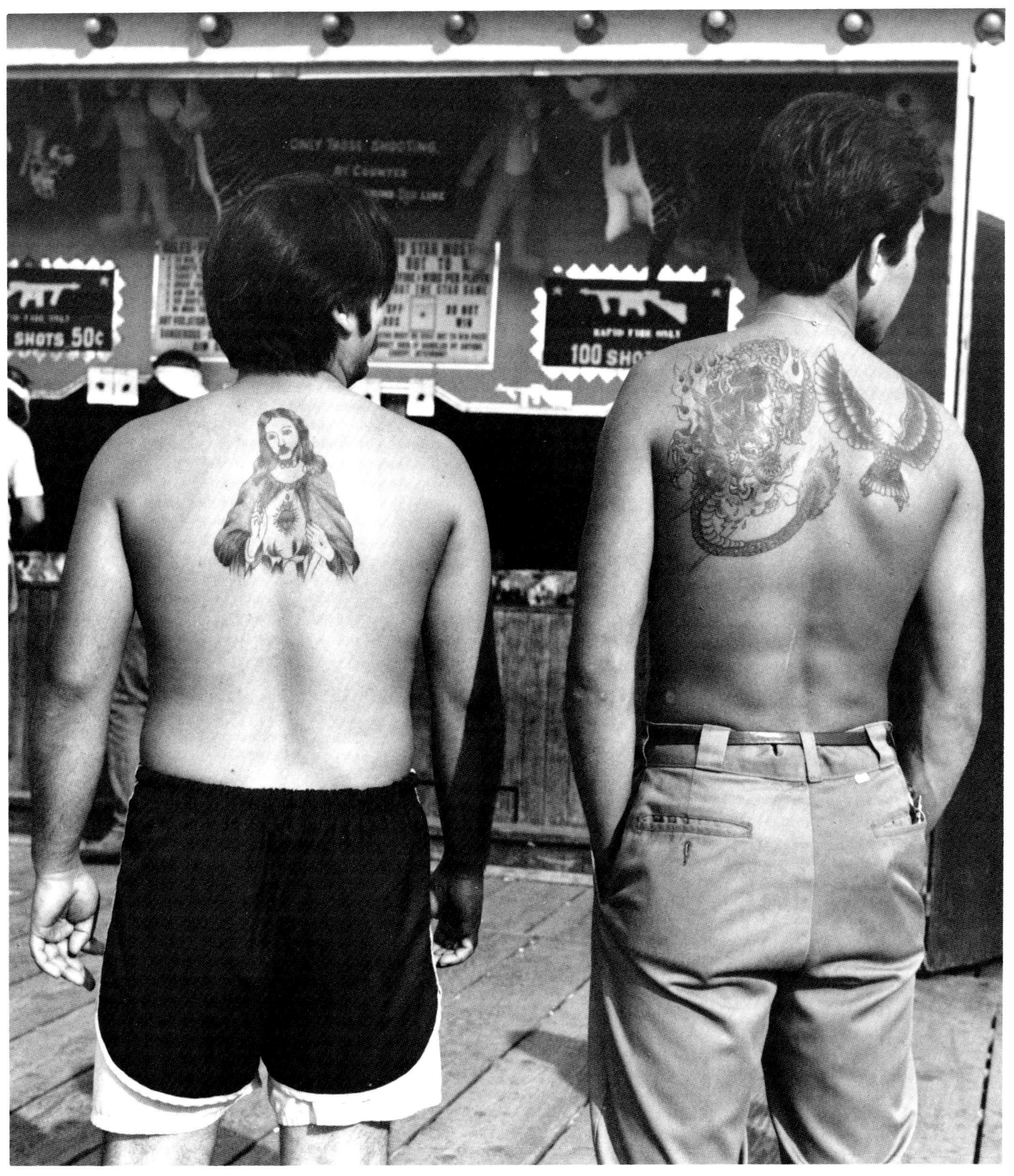

67. Danny and Victor

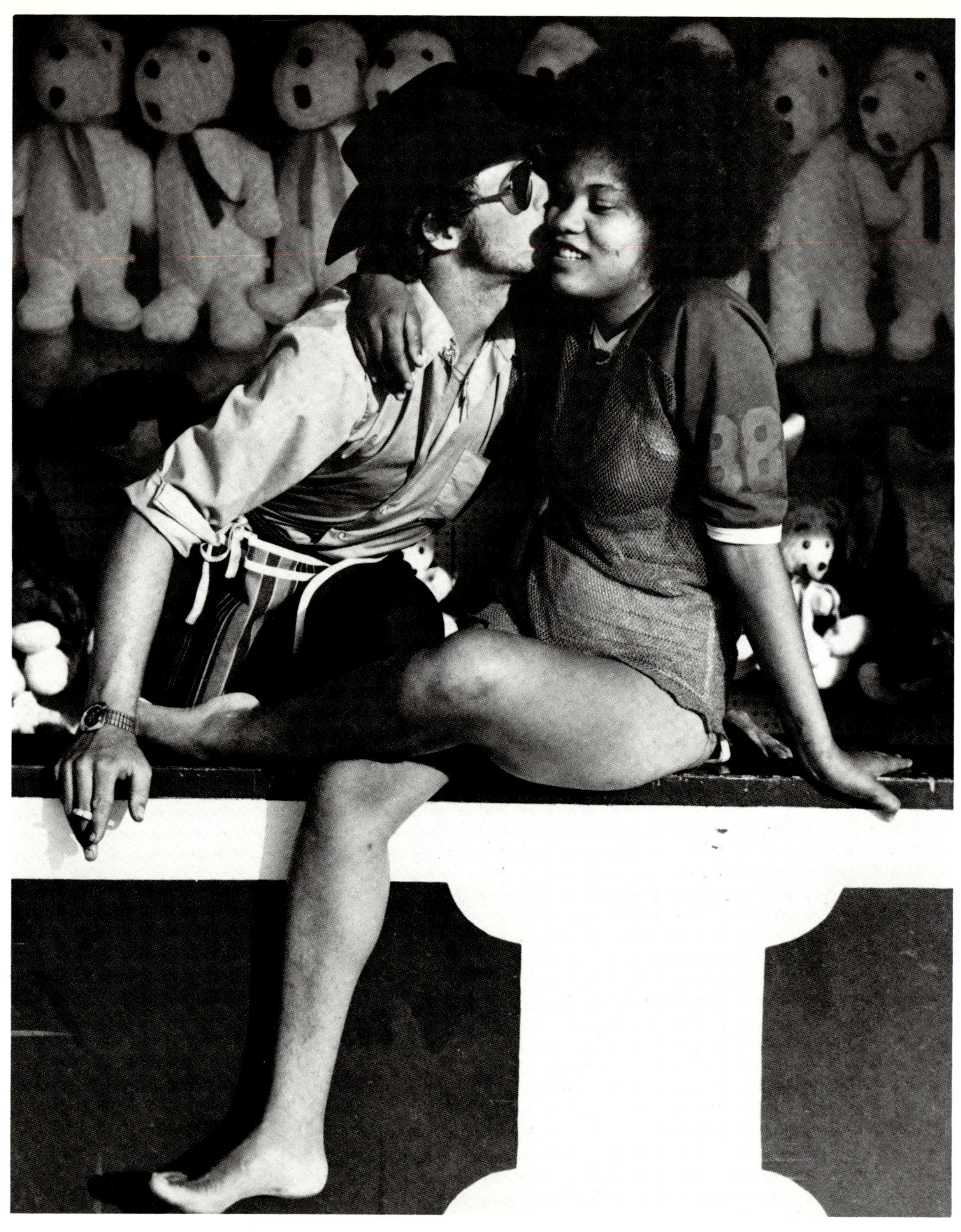

68. Cowboy and Sunshine

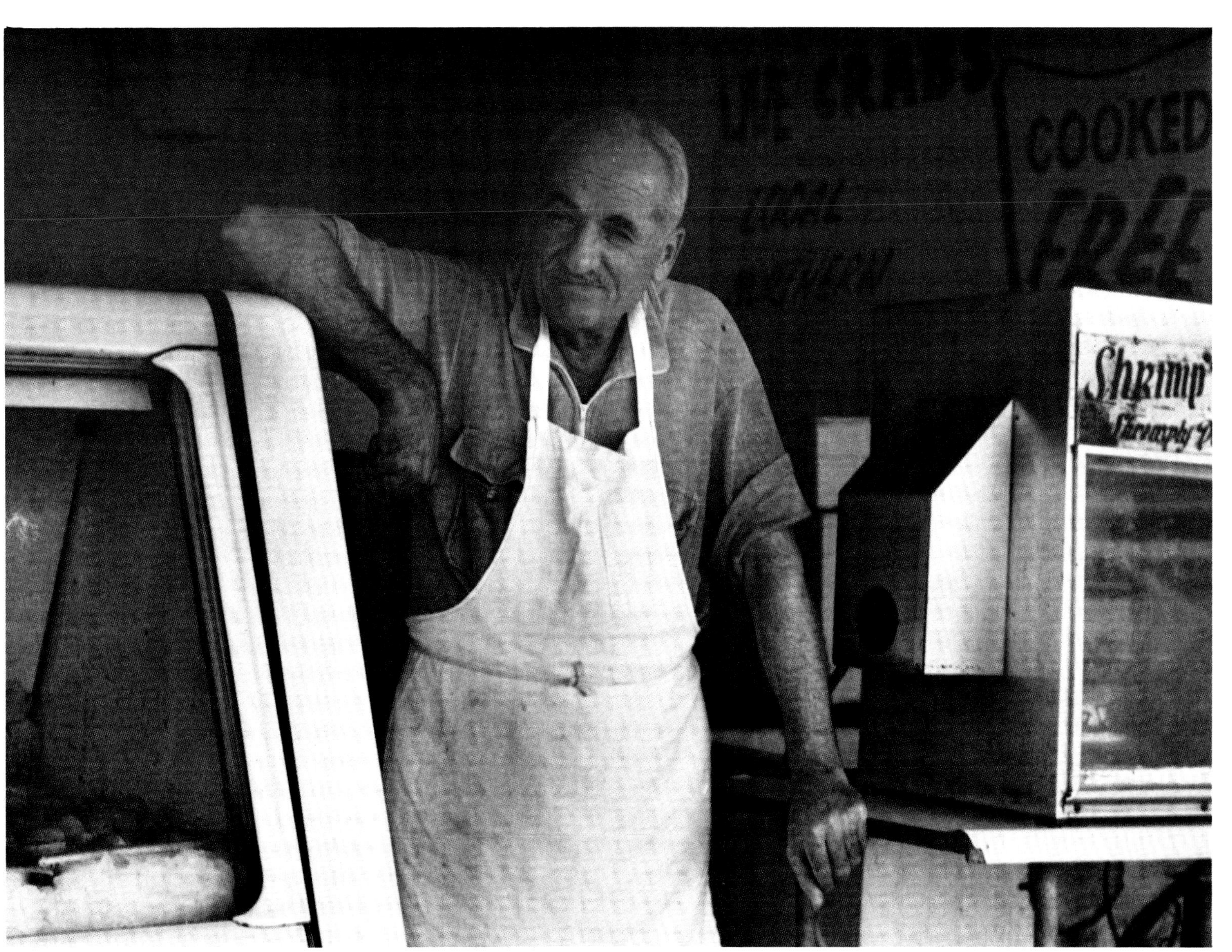

69. Frank

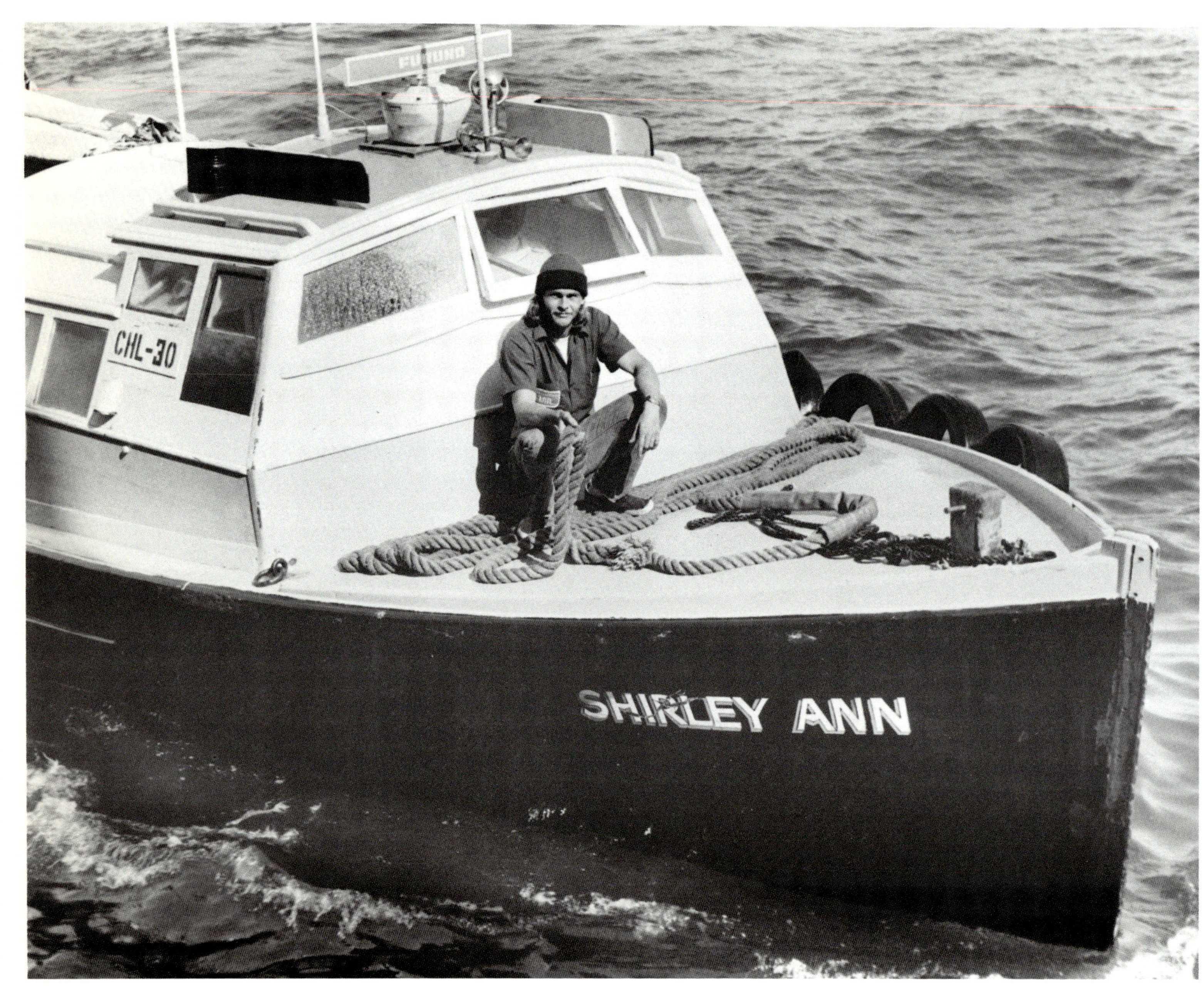

70. Water Taxi to Malibu

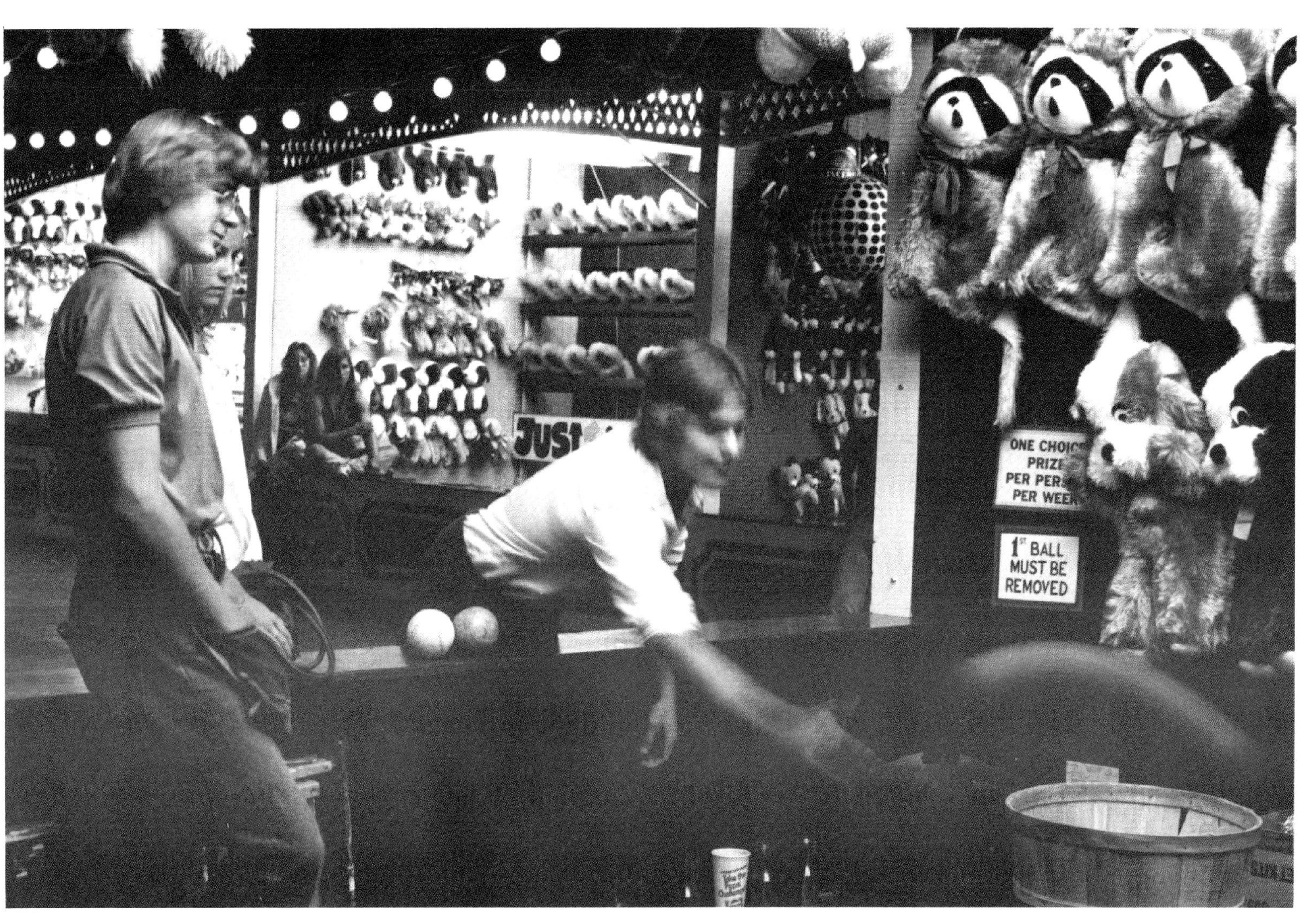

71. Ball Toss

NOTES ON THE PHOTOGRAPHS

1. Filming "Beach Patrol"—Summer, 1978
   A warm summer day. Fourth take of the same scene. The actors emote while a grip catches some rays. The scene ends, the equipment is being broken down. Brenda, a stunning blond in a very tight bathing suit, comes over and asks if I am the stills man on this picture. No, just shooting for myself. Well, do I know who's in charge? Another no. She moves off single mindedly.

2. Wall Painting, Boardwalk—Fall, 1978
   For years a wooden fence existed to keep people away from the ruins of what had been a fun zone area. This painting was on that fence. The story is that someone saved the panal with the painting. I hope this is so.

3. *Kiora* Docking—Spring, 1981
   Captain Curtis Dimmett has piloted fishing boats on Santa Monica Bay since 1938. During World War II he had to have a special Coast Guard permit and was restricted from using his ship's radio except in an emergency. He notes that fishing in the Bay was badly hurt by the eradication of sardines in the late 1960s. He fears anchovies will soon be fished out as well and there will be nothing left to draw larger fish into the Bay.

I: INTRODUCTION

4. Arcadia Hotel—1890 (Author's Collection)
   A seaside resort hotel located on the bluffs where the Pier is now. Burned down in 1909. The pier in the foreground is the ruins of the Los Angeles and Independence Railroad wharf.

5. Santa Monica Pier—1926 (Author's Collection)
   The smallest of the three Los Angeles area pleasure piers—at its height during the Roaring '20s.

6. The Santa Monica Palisades—1931 (Author's Collection)
   The view south, past the California Incline toward the Pier.

7. The Glorious 4th—1933 (Author's Collection)
   My father, Pius Zimmerman, came to Santa Monica from Richardton, North Dakota in 1932 to work for Douglas Aircraft. He and his Uncle Marcus left right after a blizzard. They arrived to find people going to the beach in the dead of winter. My father only returned to North Dakota for family visits.
8. The 4th of July—Summer, 1979
   Taken from the same spot where my father stood 46 years before.

II: THE PIER

9. Clearing Storm—Fall, 1979
   The weather closed everything except the Arcade.
10. Boardwalk Painter—Summer, 1978
   Very hot. Sipping a glass of cherry lemonade as a painter interprets the Pier.
11. Big Chicken—Winter, 1980
   Located in the Arcade. Welcoming people in the door.
12. Carousel Sign—Winter, 1980
   It took Tracey and Steve Cameron nine months to strip, clean, repair, and paint the merry-go-round animals. About a year later, the Carousel Building was closed again. This time for mechanical repairs.
13. Merry-Go-Round—Summer, 1981
   The current carousel was brought to the Pier by Walter Newcomb from the Venice Pier. It was originally built by the Philadelphia Toboggan Co. in 1922 for an amusement park in Nashville, Tennessee.
14. Westwind View of the Pier—Summer, 1978
   Skipper Loretta at the tiller. She was getting 50 cents less an hour sailing the Westwind Rental boat than she did working at the Lemonade stand on the Boardwalk. It was worth the pay cut.
15. Storm Ruins—Winter, 1983
   It does rain in Southern California. Storms too. Just down the beach the fishing boat *Patriot* was washed up on the shore. L.A.'s a harder place than you might think.

III: VISITORS

16. Monique and Kathryn—Winter, 1979
High School seniors. Came often to the Pier to skate. The sculpture they are hanging off of was in the parking lot behind the Arcade. Sand at its base. Today, Monique manages a shoe store and Kathryn helps manage the Santa Monica Mountains Wilderness Park.

17. Ouija Woman—Winter, 1979
Nodded once, agreeing to pose. Took a picture and she was gone.

18. Terri, The Seaweed Goddess—Spring, 1987
Moved down to Vista a decade ago. Mother of two and a nurse. But whenever she's back up north, she gets a sandwitch at Archie's in Culver City and comes back to the Pier.

19. Tommy—Winter, 1980
Tommy and his partner, Pather, were selling "Assa-hollah" and "Ram Iran" pins while 49 American hostages suffered in Iranian captivity. Referring to the Ayatollah Khomeini, Pather observed, "Well, it's just that he stinks, ya know?"

20. Sid, Broken Truck—Winter, 1979
28 months in the South Pacific, 34 years ago. Nightmares for years of the huge night flares fired from the Navy ships to illuminate the jungle. Now he lives in his truck and it's broken down. I said I'd never met anyone who'd been overseas that long in World War II. "You won't meet many," he said. "120 of us went over on a troopship in 1942. Weren't but 28 left to come back in 1945."

21. Laura and Snoopy—Fall, 1978
55 years old. Divorced three times. Living with an ex-junky. Wants to get out of L.A. soon. Carries Snoopy for luck.

22. Margie and Kim—Winter, 1980
Margie is Kim's grandmother. Kim loves to have her picture taken. She had to talk Margie into it.

23. George and Raul—Fall, 1979
Eating at Cocky Moon's. Both from El Salvador. Spanish station on the radio. At the other end of the Pier, another visitor, a plain 15 year old girl, asked if I wanted "to have a good time." 15. Said she had to raise money to get a place to stay. Behind and to the left, her boyfriend watched the proceedings coldly. Pity the fool who fell for this.

24. Dina and Ellen—Summer, 1982
Best friends. Just graduated high school. Headed for two different colleges in a month. Spending what time they had together.

25. Gail—Spring, 1979
All alone on the Boardwalk except for her dog, Ginger. Oklahoma accent. New to the area.

26. Singer—Spring, 1978
He usually sings down in Venice. Hot day. Hoped for big crowds at the Pier. Didn't get them.

27. Sgts. Rusty Evans and Jay Price—Summer, 1978
Recruiting for the new all volunteer army. "Ya gotta go out and find guys, now. They won't always come to you."

28. Romero—Winter, 1979
Saw me taking pictures. Told me about the Vietnam ambush that left his best friend dead and himself wounded. Took off his hat, parted his hair. A scar runs the length of his scalp. "Hey man," Romero said. "You take my picture. Call it 'The Vietnam Vet—I went off to fight for my country and no one gives a fuck.'"

29. Jerimiah and Diana—Summer, 1979
Jerimiah just returned from Australia. He is an Anthropologist.

30. Carla—Spring, 1984
Hard wind off the Bay. The jacket is too big.

31. Mid-Winter Tourists—Winter, 1980
Sunshine and no snow. They were happy to be here.

32. Utzer—Summer, 1979
Kit Utz is a Westside real estate salesman. Runs on the beach every day. Sometimes he plays the games. Sometimes he wins.

33. Chris—Fall, 1979
Holloween celebration on the Pier. Lucky the Clown was down by the Carousel, entertaining children from the Santa Monica school system.

34. Carol and Caterina with "Tom Cruise"—Fall, 1988
In Washington, D.C. they have cut outs of the President.

35. Lupe—Fall, 1978
On the fishing platform at the end of the Pier. Beer in a brown bag. Leaning against the back wall, while in front of her dozens of guys—including her boyfriend—try to catch their limit.

36. Helicopter Ride, Arcade—Summer, 1982
    For children only. Adults wishing to relive the controlled fears of childhood have to look elsewhere. Someone over 75 pounds climbs on and the plug gets pulled.

37. Adrian and Elizabeth—Fall, 1979
    Halloween. In front of Doreena's, the Fortune Teller. Doreena and her sister, Sandy, took over the psychic business from their mother. She was a Rumanian who hung out her shingle near the Ocean Park Pier in 1943. Sandy said they are happy to give advice on "whatever somebody is interested in."

38. Jeannie—Winter, 1990
    L.A. winter. Under the Pier.

39. Fun Zone—Spring, 1988
    The Carousel Park Playground, simple rides aimed at kids, was one of the additions to the Pier as it was gradually repaired. The Fun Zone is opened in the Spring and Summer.

40. The Punching Bag—Summer, 1983
    Arcades started out as male preserves. This is changing, of course, but lots of the games still reflect their origin. The harder you hit this bag, the more clothes the woman illustrated on the machine takes off. Pin ball games inevitably featured Frazetta women sporting nipples stiff enough to hang your hat or coat on.

IV: REGULARS

41. Yvonne—Summer, 1981
    Yvonne sells the tickets to the bumper car ride. Her daughter, Susan, worked in one of the carny games which were all run by Susan's uncle, Bob.

42. In the Parking Lot—Spring, 1982
    A group of regulars—at home.

43. Jay and Cissy, Wayne, George and Dawn—Summer, 1979
    On the Municipal Pier, over the Pacific.

44. Ron Griego—Summer, 1978
    Jumped over four trash cans in roller skates. For sport. No hat was passed.

45. On the Boardwalk—Summer, 1978
I swear this wasn't set up.

46. Slim Jim and Virginia—Winter, 1980
Virginia comes down to the Pier from Hollywood. Slim Jim is originally from Platsburg, New York. He thumbed his way out West "to see the sun set over the ocean." He notes that when trouble happens at the Pier it comes from outsiders, not the people who hang out there. Jim works construction when he's not at the Pier.

47. Crispy Cory—Summer, 1981
When I asked about him five years later, Cory was married with two daughters.

48. Mary and Beauty—Spring, 1979
In the parking lot off the Boardwalk. Mary was at the Pier all the time. I never saw her friend, Beauty, again.

49. Laura—Summer, 1981
Laura ran the dime toss. Throw a dime and win a plate, glass, or other fine prize. After the summer she went back home to El Paso. After graduating from Bowie High School and college, she joined the Army and is now Lt. Laura Lopez.

50. Sean, Jerry, Randy, Sandy, and Mac—Spring, 1979
At the Muscle Inn on the Boardwalk. Sandy was brought over for the photo, then returned.

51. Lolita—Summer, 1978
"Are you taking pictures of me?" she asked. Sure am. She smiles. I turn around to see Humbert Humbert standing behind me, looking on knowingly.

52. On Jerry's Bike—Summer, 1978
Jerry saw me passing by. "Hey man, take a picture of my bike and my old lady." Sure. She lay back on the bike. "Look sexy," Jerry says. "Think of an orgasm."

53. Crispy Cory, After the Accident—Spring, 1979
Cory spilled his bike on a freeway off ramp. Lacerated stomach, two broken ribs. Not the kind of guy you'd even ask if this will make him give up on motorcyles.

54. Lower Pier—Summer, 1982
This is the landing where fishermen boarded the *Kiora.* It was also a good place to be alone. Swept away in the storms of 1983.

55. Don—Fall, 1979
Don had once written a novel about the mines near his home in Redding, Calif. But the manuscript was stolen along with his clothes and $60. So he had to sell his rod to raise some cash. Later he became a security guard on the Pier.

56. Angel and Aaron—Fall, 1978
Next to Beryle's Art Novelties where you could choose from a truly staggering variety of plaster objects.

57. Charlie and Lisa—Summer, 1979
On the Boardwalk.

58. May McCabe—Summer, 1982
May runs the Arcade. She came to the Santa Monica Pier in 1965 after working at Pacific Ocean Park for years. During World War II she worked at a spaghetti house in Ocean Park and recalls the barage ballons and anti-sub net around the Pier—which was inevitably knee deep in soldiers and sailors.

59. Al the Photographer—Summer, 1979
Al made his living selling Polaroid pictures on the Pier and elsewhere. Normally, they were bought by working class people who could afford his pictures but not a good camera.

60. Doug—Summer, 1978
At the Hamburger Haven on the Boardwalk.

61. The Beating—Summer, 1979
The fellow on the bottom had pushed around a ten year old kid who then went and told his friends. The shover was one of those vacant men. After the beating, the police arrived and arrested him. As the other guys gave their statement, the cuffed man sat in the police cruiser, staring out the window with no sense of surprise.

62. Wild Bill—Summer, 1981
Bill was originally from Boston where he was in charge of typewriter repair at Harvard University. He came out West because he wanted the sun. He worked at Sea Mist Rentals, repairing and renting out roller skates.

63. Nancy, Critter, and Matina—Fall, 1980
A cold day. Critter had been incognito for awhile for reasons he chose not to disclose. He was glad to be back.

64. Tasha and the Manager—Summer, 1978
Tasha was running the Coke ring toss this night. She was paid $2.65 an hour plus a percentage of the take.

65. Mario—Summer, 1981
Mario Torres was born in the Bronx. After a stretch in the Marines, he and his wife, Lisa, settled in Santa Monica because "New York is too fast." Mario always ran the Water Balloon Race, received a large percentage of the business, and did quite well for himself. He died in a car accident four years after this picture was taken.

66. Man in a Storm—Winter, 1979
He lit a cigarette and went over to the Arcade, the only thing opened on this blustery day.

67. Danny and Victor—Summer, 1979
They were with two other guys. Referred to themselves as "the Westside homeboys."

68. Cowboy and Sunshine—Summer, 1979
Sunshine liked the idea of a photo series on the Pier, but thought I should change the name. She suggested "Pier Passion" because "it's sexier."

69. Frank—Fall, 1978
Frank ran Sea Food Specialties next to Beryle's Art Novelties. He sold fresh fish caught in the Bay and elsewhere.

70. Water Taxi to Malibu—Spring, 1979
A rock slide closed the Pacific Coast Highway. The only way to get between Santa Monica and Malibu was to take this ferry between the two Piers. It cost $2.00 for a one hour voyage.

71. Ball Toss—Summer, 1978
Summer nights. The Pier's open late.

BIBLIOGRAPHY

Barry Brennan, "Wednesday People Fill Camera's Lens," *Santa Monica Outlook*, October 18, 1982, p.A7.

Raymond Chandler, *Farewell My Lovely* (New York: Knopf, 1940).

Jean Femling, *Great Piers of California: A Guided Tour* (Santa Barbara: Capra, 1984).

Mark Foster, "Santa Monica Goal: Showplace in Pier Face Lift," *Los Angeles Times*, August 17, 1978, p.VII1+.

Bruce Henstell, *Sunshine and Wealth: Los Angeles in the '20s and '30s* (San Francisco: Chronicle, 1984).

Molly Jensen, *Time Out: Sports and Leisure in America Today* (Tampa: The Tampa Museum, 1983).

Sam Hall Kaplan, "The Pier Pressure in Santa Monica," *Los Angeles Times*, September 13, 1982, p.V1+.

Gary Kyriazi, *The Great American Amusement Parks* (Secaucus: Citadel Press, 1976).

Ernest Marquez, *Port Los Angeles: A Phenomenon of the Railroad Era* (San Marino: Golden West, 1975).

Charles Moore, Peter Beeker, Regula Campbell,*The City Observed: Los Angeles* (New York: Vintage, 1984).

Roberta Ostroff, "Symbols of Seaside Fun: Aging Piers Live On," *Los Angeles Times*, July 22, 1979, p.VII1 + .

Gillian Rees, "Last of the Peerless Piers," *Westways*, v.74 (June, 1982), p.29-31 + .

Bill Rollins, "Tom Zimmerman's Pier People," *Los Angeles Times*, August 28, 1983, p. E1-2 + ; and, September 25, 1983, p. E1-2 + .

Jeffrey Stanton, *Santa Monica Pier: A History from 1875-1990* (Los Angeles: Donahue, 1990).

—*Venice of America: Coney Island of the Pacific* (Los Angeles: Donahue, 1988).

Palmina Stephens, "Giving New Life to a Carrousel," *Los Angeles Times*, June 1, 1981, P.V1 + .

ABOUT THE AUTHOR

Tom Zimmerman is a native of Los Angeles and has been visiting the Pier his whole life. His word and photo essays have appeared in a variety of publications, including *The Los Angeles Times Magazine*, *Americana*, *Conde Nast Traveler*, *Southern California Quarterly*, and *Journal of the West*. His previous book, *A Day in the Season of the L.A. Dodgers*, was published by Shapolsky in 1990. His photographs have been exhibited across the country, and are in the permanent collections of the Brooklyn Museum, National Baseball Hall of Fame, Director's Guild, California State Library, Los Angeles Library, and the Historic American Building Survey at the National Archives.